Rumors of Existence

For Corinne Elizabeth Bille,
for whose generation we must explore
and preserve the wonders of Creation,
And for Deb, my partner in all life's endeavors.

Rumors of Existence

Newly Discovered, Supposedly Extinct, and Unconfirmed Inhabitants of the Animal Kingdom

Matthew A. Bille

hancock

house

ISBN 0-88839-335-0
Copyright © 1995 Matthew A. Bille

Cataloging in Publication Data
Bille, Matthew A.
    Rumors of existence

    ISBN 0-88839-335-0

    1. Rare animals.   I. Title.
QL706.8.B54 1995   599'.0042   C95-910640-5

Cover painting and design: Karen Whitman
Editor: Karen Kloeble
Production: Nancy Kerr and Myron Shutty
Cover painting: Tasmanian tiger
Back cover photo: author with a Cuban ground iguana

Published simultaneously in Canada and the United States by

HANCOCK HOUSE PUBLISHERS LTD.
19313 Zero Avenue, Surrey, B.C. V4P 1M7
(604) 538-1114   Fax (604) 538-2262

HANCOCK HOUSE PUBLISHERS
1431 Harrison Avenue, Blaine, WA 98230-5005
(604) 538-1114   Fax (604) 538-2262

Contents

Preface

You are about to meet some of the least-known animals in the world.

Despite all the scientific progress of the twentieth century, some creatures remain in the shadowlands of zoology. These are the animals just being discovered, the presumed-extinct animals that may not really be gone, and the mysterious creatures which are not yet recognized as "official" inhabitants of the animal kingdom.

The first section of this book includes the newest finds: animals entered into the reference books in the last sixty years. That time span was chosen because it takes us back to the 1930s, when science first met some important and surprising animals which deserve inclusion in a book like this.

Section II introduces animals usually thought of as extinct which may still be clinging to existence. Also recounted here are the stories of creatures whose extinction was considered a fact, but which nonetheless made a reappearance.

In Section III, we meet the mystery animals. This section is not about alleged "monsters" like the sasquatch, but about creatures which remain unrecognized even though physical remains are in hand or a qualified scientist has met a specimen closeup in the wild. You may be surprised at how many such cases there are.

Every animal discovered or rediscovered is an expression of the diversity of this world, and a plea to preserve it. If we continue to burn, bulldoze, and driftnet, many of the rarest species will disappear, and some fascinating discoveries will be lost without our ever knowing they were there.

Fortunately, the finding of new or supposedly extinct animals also gains wide public and scientific attention. The search for such animals may play a crucial role in convincing people that Earth's vanishing wilds deserve to be saved.

Acknowledgements

Writers, even more than a certain fictional character, depend on the kindness of strangers. Many people have been kind to me in the production of this book, and some, thankfully, are strangers no longer.

The greatest appreciation is due to Craig Gosling, who enthusiastically employed his artistic talents to make the creatures in these pages come to life. Zoologist Dr. David Daniell of Butler University also made a crucial contribution by proofing the manuscript for scientific accuracy.

Professor J. Richard Greenwell, secretary of the International Society of Cryptozoology, was generous with his time, as was Dr. Roy Mackal. The ISC's periodicals offered a very useful source of information on newly discovered or rediscovered animals. Dr. Karl P. N. Shuker, British zoologist and writer, supplied research material and patiently answered my queries concerning his work on the peculiar Kellas cat. Drs. James Halfpenny, Robert Ridgeley of the Academy of Natural Sciences, James Platz of Creighton University, and Troy Best of Auburn also shared their expertise or provided material, as did French researcher Michel Raynal and British biologist Michael Bright. Special appreciation is due to my mother, Jane Bille, for translations of French-language material. John and Linda Lutz of the Eastern Puma Research Network contributed their data on reports of that animal.

The staff of the Worldwide Fund for Nature was very helpful, as was Jill McLaughlin of Conservation International, and Alan Rabinowitz of the Wildlife Conservation Society/NYZS. These people and organizations loaned invaluable pictures in addition to providing information. The staff of the Kokomo, Indiana public library assiduously tracked down obscure books through interlibrary loan. Suzanne Braun of the Indianapolis Zoo library was just as helpful.

Any writer who is a parent owes a debt to his child's caregivers, and so I thank Tricia Childers and Michelle Myers. I thank Tricia for the author photograph as well, and I appreciate the assistance of public relations officer Meg Beasley and reptile curator Bill Christie of the Indianapolis Zoo for arranging the setting. Finally, I need to thank the family and friends who encouraged me to keep writing: my parents, Deb's parents, Liz Ruth, Gregg Anderson, and many others.

An Unavoidable Note on Classification

Animals often appear in or disappear from scientific literature based, not on new findings, but on revised classifications. This book tries to avoid such technical matters of taxonomy, but sometimes the subject just can't be dodged. Accordingly, some readers may want a brief review. The world of living things is divided into kingdoms. Originally there were just two: plants and animals. The current consensus is that there are five, with separate kingdoms for the bacteria, the fungi, and the protists (those microbes that don't fit anywhere else).

This book is limited to the animal kingdom, which includes many subdivisions. The major ones, using humanity as our example, are phylum (Chordata), class (Mammalia), order (Primates), family (Hominidae), genus (Homo) and species (sapiens). All these may be broadened or narrowed using the prefixes super- and sub-, creating, for instance, superspecies and subspecies. As we are the "type" subspecies of *Homo sapiens*, it is proper (if clumsy) to refer to modern humans as *Homo sapiens sapiens*.

The basic building block of classification is the species. In general, a species is a population of animals that *normally* interbreeds with each other and not with anything else.

This doesn't mean there is one agreed-upon list of species. The genus *Aotus* (night monkeys) includes anywhere from two to nine species, depending on which opinion you accept. There are disagreements about classifying almost *all* of the forty-odd species of dolphins, producing a taxonomic chop suey that is frustrating to experts and nearly incomprehensible to anyone else.

Differences within a species may be recognized by dividing it into subspecies, although the delineation of subspecies is an imperfect art at best. Two other divisions sometimes used are "variety" and "geographic race." Some writers use these labels interchangeably with "subspecies," while others give them their own meanings. This is confusing even to zoologists, some of whom now advocate giving up on formal names below the species level. This book deals mainly with species, but well-defined subspecies will be mentioned on occasion.

A final factor is that, until the early twentieth century, scientists tended to be "splitters," creating new species based on very minor differences. When this led to eighty-six species of the brown bear, it became obvious things were getting out of hand. Today's zoologists are mostly "lumpers," and have reduced that swarm of bears to a single species with just four subspecies.

The subject is immensely more complicated than this, but elaborating here would just obscure our topic. Accordingly, we'll stop classifying animals and press on to meeting them.

Section I

Recent Discoveries

Introduction: The Newest Animals

Most animal discoveries don't make news. New invertebrates or mice or lizards may excite the scientific community, but they rarely draw much interest from the media. Accordingly, many people don't realize that new animals, large and small, are still being found. For example, the years since 1940 have seen an average of three new birds and more than three new mammal species described annually.

Here are just a few of the recent discoveries, some of which will be revisited in more detail later in this section.

We will start in 1937, the year the latest really massive land animal was found. This was the kouprey, a wild ox from Southeast Asia. A year later came the most famous of "living fossils," the ancient, lobe-finned fish called the coelacanth.

The world's most primitive wild cat, the yameneko, was found on the Japanese island of Iriomote in 1965. A 1975 expedition to Paraguay encountered herds of the Chacoan peccary, a 100-pound wild pig believed extinct since the Ice Age. The Peruvian beaked whale was discovered when a single skull was found washed ashore in 1976. A new albatross appeared in 1983, and a previously un-known octopus with a span of eight feet was filmed by French oceanographers in 1984.

1985 brought the discovery of the Sanje mangabey, a new mon-key, from Tanzania. In 1990, another monkey, the strikingly colored black-faced lion tamarin, turned up for the first time in the heavily populated Brazilian state of Parana.

Two years later, scientists in Vietnam announced the amazing discovery of a large, hoofed mammal. The Vu Quang ox appears to be an important clue to the evolution of all cattle. Following on the hooves of this discovery came evidence of *at least four more sizable mammals* from the same region. It is impossible to overstate the impact of this unforeseen and almost unprecedented zoological gold rush.

The discovery of new species may have slowed a bit in the twentieth century. But the surprises keep coming, and it would be foolish to say there will be no more.

The Elusive Beaked Whales

The whales should have been completely cataloged many years ago, or so one would think. After all, these mammals are generally large and must come to the surface periodically to spout, announcing their presence in a very conspicuous fashion. Moreover, they have been hunted relentlessly for centuries.

Even so, some members of the order Cetacea have been rare enough, shy enough, or lucky enough to avoid the gaze of humans until recently.

The least-known cetaceans belong to the reclusive, deep-diving family known as the beaked whales. Beaked whales have dark upper bodies (which may be black, gray, bluish, or brown) with pale undersides. They are distinguished by long, dolphin-like snouts, which in most species contain only two functional teeth. The placement and shape of these teeth are important criteria for classifying the different species.

The Indopacific beaked whale, whose teeth are in the very tip of the lower jaw, is known only from two skulls washed ashore thousands of miles and 135 years apart. There are no known sightings of the whole animal, alive or dead, and we know absolutely nothing about it except the bare facts that one skull was found on an Australian beach in 1822 (it wasn't formally described until 104 years later!), and that a second skull was found in Somalia in 1955.

The Tasman whale, also called Shepherd's beaked whale, was unheard of until a stranding in 1933. A dozen subsequent strandings since then have given us all our meager data on this whale, which reaches well over twenty feet in length. One possible sighting from New Zealand in 1964 represents the only reported encounter with the living animal. Unlike other beaked whales, Shepherd's has a full set of small functional teeth in addition to the two major ones.

Andrews' beaked whale is known only from remains washed up on South Pacific shores. Another beaked whale, Hector's, was described from a few strandings. A photograph taken off Catalina Island in 1976 apparently documents the first live sighting ever.

The ginkgo-toothed whale, up to sixteen feet long and weighing over 3,000 pounds, was first encountered in 1957. The name of this black-and-gray denizen of the north Pacific derives from the unusual leaf-like shape of the two teeth. The story of its discovery is unique, to say the least. The first known example wandered into shallow water off Tokyo's Oiso Beach. Japanese boys playing baseball on the sand waded into the water and beat the hapless animal to death with their bats.

The stoutly built arch-beaked whale, or Hubbs' beaked whale, was introduced to science when a specimen washed ashore in Washington State on November 2, 1944. After a long period in which it was confused with the species known as Stejneger's beaked whale, the arch-beaked was finally identified by Dr. Joseph Moore in 1963. This whale's beak is tipped with white, and its two six-inch teeth are prominently exposed even when the jaws are closed.

The smallest and most recently discovered beaked whale is the Peruvian. Scientists had no inkling of its existence until 1976, when Dr. James Mead found its decaying skull on a beach in Peru. Mead realized the skull was from an unknown type of whale, and he immediately began a search for better specimens. This took quite a while. It is the adult male whose teeth are used to distinguish one beaked whale from another (in the females, the teeth often don't erupt from the gums at all), and it wasn't until 1988 that Mead obtained such a specimen.

By the time Mead formally published his description of the species in 1991, Peruvian scientists and fishermen had helped him assemble a total of eleven specimens. All were found either washed up on shore or trapped in fishing nets.

The largest example was only twelve feet long, so it's not surprising that the Peruvian is also referred to as the pygmy beaked whale. There have been no scientific observations of the living whale, so its range, habits, and population size remain unknown.

Leaving the beaked whales, we come to the case of another "pygmy." Whalers in the Antarctic were catching small blue whales ("small" meaning just over the seventy-foot legal minimum length set by the International Whaling Commission) for decades without thinking much about it. Dr. Tadayoshi Ichihara of Japan, who examined the blues taken by his country's whaling fleet, was the first to notice the number of fully mature whales in this size range.

In 1963, Ichihara formally described the pygmy blue whale as a separate animal. He named it *Balaenoptera musculus brevicauda,* thus classifying it as a subspecies of *Balaenoptera musculus,* the standard blue.

This discovery was initially controversial. Some conservationists alleged the description was not scientific, but economic, charging that Japan's whaling industry wanted the pygmy blue recognized so the International Whaling Commission would lower its minimum size limit for blues. Dr. Ichihara stood his ground, arguing that pygmies showed unique characteristics other than size. The pygmy has a proportionately longer trunk and smaller tail section than its giant kin, and its color is mainly silver-gray instead of the normal deep blue or gray-blue.

According to the authoritative *Walker's Mammals of the World,* most experts now agree the pygmy blue is a valid subspecies. At least one cetologist, noting that the blue and its pygmy kin have partially overlapping ranges but apparently do not interbreed, has even suggested the pygmy is a species in its own right.

Whales are among the most intelligent and fascinating creatures of the world. It's both exciting and a little humbling to think we still have a lot to learn about them.

The Yameneko

In 1965, Japanese naturalist Yukio Tagawa traveled to the Pacific island of Iriomote, a mountainous, forested dot of land in the Ryukyu chain. His purpose was to investigate reports of a wild cat on the island. His discovery—the yameneko, or Iriomote cat—was a stunner to zoologists. No one had described a new cat since 1892, when the Chinese desert cat was discovered.

When the first specimens reached Tokyo, Dr. Yoshimori Imaizumi set about the task of classifying the animal. Its closest relative appeared to be the leopard cat, which ranges across Asia and a number of Pacific islands. The yameneko was sufficiently unique, in Imaizumi's view, to warrant creating a new genus.

Most zoologists disagreed on that point, assigning the new cat to the genus *Felis,* which houses all the "small cats" and the puma. Still, everyone found the Iriomote cat very interesting. Dr. Imaizumi believes it has lived in isolation, without undergoing any significant evolution, for some three million years. That would make it the oldest cat species in the world today.

In general size and appearance, the yameneko resembles a stocky, short-legged house cat. Its fur is mainly brown, with darker spots that form irregular bands along its sides. The ears are round and marked with white stripes. The cat's primitive features include claws that don't quite retract all the way and unique external scent glands under the tail.

As befits a cat, the yameneko is a shy creature which remains mysterious. Very little is known about its habits, and no one is sure when or how it got to Iriomote in the first place. The island's natives had always been aware of the cat, but knew little about it except that it was good to eat.

The yameneko is now protected by law, but it remains endangered. Its forest habitat is shrinking, and there may be less than fifty of the animals remaining.

A 1974 Japanese-German expedition to study the cat in the wild produced an unexpected bonus. Among the yameneko's prey species was a dwarf pig, also new to science!

At this writing, the yameneko is the last full species of cat to be discovered. A new species provisionally announced from the Japanese island of Tsushima in 1989 is now thought to be a subspecies of the leopard cat, although it still ranks as a significant discovery.

If rare cats like these are to survive, they will need all their proverbial nine lives—plus strict conservation laws and a generous measure of luck.

Megamouth

Zoologists are always pleased when a new species turns up, however small and inconspicuous it might be. So they were absolutely thrilled in 1976 when a Navy vessel cruising off Hawaii came up with a weird-looking and totally unknown fifteen-foot shark tangled in its sea anchor.

The megamouth shark proved to be not only a new species but the sole member (so far as we know) of a new family. The shark is a harmless filter feeder with an oversized head and a gaping mouth which turns down sharply at the corners, giving it a perpetual grimace. It's a chunky, slow-swimming, shy creature, hardly fitting the usual picture of a shark at all.

It was eight years before a second specimen was netted. Since then, four more have turned up, including one in Australia and two in Japan. The latest was caught alive off California in October, 1990. The 1200-pound fish was measured (sixteen feet, three inches), photographed, and finally released with two miniature transmitters embedded under its skin.

These revealed the animal is a vertical migrator, rising and falling every day in conjunction with its food, a layer of tiny animals called zooplankton. The creatures making up this diet are lured into the shark's four-foot-wide mouth by a bioluminescent lining and strained out by 236 rows of very small teeth.

At night the shark remains at a relatively shallow depth of about forty feet. At dawn, it descends to 500 feet or deeper and stays there all day. This striking behavior—unknown in any large fish until now—undoubtedly played a role in the megamouth's avoiding human contact for so long.

Megamouth may be the most surprising shark discovery of recent times, but it's far from being the only one.

Herman Melville offered a typical nineteenth-century view of the shark as "...the dotard lethargic and dull, pale ravener of horrible meat." A more modern appraisal is that the sharks are an ancient and highly successful order, having adapted to fill every kind of ecological niche in the vast oceans. New species, large and small, are still being added to the 300-plus sharks known to humans. Most of the recent finds are deepwater predators who rarely cross paths with science.

The Pacific sleeper shark, discovered only in 1944, is a particularly large and particularly deep-dwelling example. How large is it? The biggest one ever captured was thirteen feet long, but in 1989, the submersible *Nautile* was diving in Japan's Suruga Bay when, as one researcher put it, "We saw a fish bump into a wall, and then the wall moved. The sub shook." Sliding past the *Nautile* was an awesome sleeper shark, at least twenty-three feet long.

Pacific sleepers cruise the darkness as far south as California, chomping on a typically varied shark diet including squid, octopus, fish, crabs, and anything else smaller than themselves.

Other deepwater families include the sixgill sharks (most sharks have five pairs of gill slits). The bigeye sixgill, the second species of sixgill shark known, was described from the Bahamas in 1969. It has since been reported from Florida, the Philippine Islands, and the Indian Ocean, indicating a wide-ranging distribution in its chosen waters 120 to 220 fathoms beneath the surface. The bigeye sports a common shark color pattern, gray on the back and lighter underneath, and can reach six feet in length.

A cruise through other modern shark discoveries includes such distinctive types as Australia's maneating bronze whaler, up to four-teen feet long and first identified in 1938. The silky shark, a ten-foot species named for its smooth skin, was classified in 1943 and turned out to be numerous in both the Atlantic and Pacific oceans. A Pacific-only species, a salmon shark reaching at least eight feet in length, joined the official lists in 1947.

A shark known only from depths of 2000 feet or greater, the twelve-foot bigeye sandtiger, was encountered for the first time in 1955. The long-finned mako shark was described in 1966.

A rare kind of thresher shark, the third member of its genus, was described by a Japanese ichthyologist in 1935, forgotten about, and then confirmed in 1975. It has the exaggerated tail lobe of the other thresher sharks, but smaller teeth, among other distinguishing characteristics. The bottom-dwelling sailback houndshark, six feet long, was discovered in the South Pacific in 1973.

A thirty-one-inch sawshark of a new species, *Pristiophorus schroederi,* was caught off the Bahamas in 1959. The sawsharks resemble the better-known sawfish, but are smaller and are true sharks, whereas the sawfish are actually rays. The angelsharks are another raylike group. The Taiwan angelshark is known from a lone immature specimen caught in 1972, and the ocellated angelshark from a single catch in the same region ten years earlier.

The largetooth cookie-cutter shark was described in 1964. Less than a foot and a half long, it has a projecting nose and straight back which make it look remarkably like a German U-boat without the conning tower. Cookie-cutters are strange little sharks equipped with outsized teeth to cut circular "plugs" of flesh from whales and larger fish. The largetooth has been found off Okinawa and in the Gulf of Mexico. As with many deepwater species, we know almost nothing else about it.

All sharks have their special adaptations. The taillight shark was first caught off Uruguay in 1966. Its name is no misnomer: a gland below the tail of this three-foot shark produces a secretion giving off a bluish glow.

At the opposite end of the size scale from the sleeper shark are the dogsharks (or dogfish), a large family of mainly diminutive species. In 1985, a new dwarf dogshark was described after being caught a thousand feet down in Carribean waters. At seven inches long, it is among the smallest sharks known (we simply don't know enough about sharks to declare which species is definitely the smallest). Another new dogshark from the same area is less than nine inches long.

There are so-called catsharks in the deep waters, too: the harlequin catshark was caught for the first and only time in 1973. The campeche catshark was first caught in 1979, the South China catshark in 1981, and the spotless catshark in 1982.

These are just a few of the sharks encountered for the first time in recent decades. There seems to be no end to the variety of this ruling clan of fishes.

The Coelacanth Mystery

There is no way to write a book about recent zoological discoveries without paying tribute to the coelacanth. When this heavily built fish swam out of the age of reptiles into a South African trawler's net in 1938, it proved that descendants of a prehistoric species could exist undiscovered long after its fossil record had come to an apparent end. Credit for this epochal find goes mainly to Marjorie Courtney-Latimer, the East London Museum curator who first recognized the trawler had brought in something odd, and Professor James L. B. Smith, the ichthyologist who was stunned when he realized just what that "something" was.

Technically, we're talking about "*a* coelacanth": the living species, *Latimeria chalumnae,* is the only known survivor of the ancient order *Coelacanthini.* However, no one except an ichthyologist is ever going to refer to the fish as anything other than "*the* coelacanth."

This big blue antique boasts a number of peculiar features, such as a joint in the top of the skull and a soft, hollow notochord filled with oil in place of a true backbone. But the most attention has been focused on its pectoral fins, which look like stumpy, fringed legs. In fact, the fishermen who caught the first specimen, having no idea what the five-foot-long creature's proper name was, christened it "the great sea lizard."

The coelacanth is a survivor of the lobe-finned fishes which, it is believed, gave rise to the amphibians, which in turn evolved into the reptiles, mammals, and birds. The living species is not considered our direct ancestor, but it still has a lot to tell us about the evolution of all the vertebrates.

Poet Ogden Nash celebrated the coelacanth's stubborn refusal to evolve with typical wit:

It jeers at fish unfossilized
As intellectual snobs elite;
Old Coelacanth, so unrevised,
It doesn't know it's obsolete.

Since that first amazing catch, over a hundred additional specimens have been obtained. (In fact, there is some fear that the coelacanth may be driven back into extinction by the quest for scientific knowledge.) Professor Hans Fricke led a German expedition in 1987 which used a submersible to film the coelacanth in its natural habitat.

That habitat is generally believed to be limited to the Comoros Islands and the surrounding area off the coast of Madagascar. Oddly, the very first coelacanth was caught some eighteen hundred miles away, off South Africa. No South African specimens have been caught since, so that exception is considered a stray. There was, however, a 1991 catch off Mozambique. Is this another stray, or is the coelacanth's range wider than we thought?

There is an even stranger final chapter to this scientific mystery novel, and the last page may not have been written. In 1949, a souvenir shop owner in Tampa, Florida, purchased a bucket of what she thought were tarpon scales from a local fisherman. (The scales were used in making nautical knick-knacks.) On closer examination, the shopkeeper decided these scales were of a type she didn't recognize. She mailed one to Washington, D.C., where Dr. Isaac Ginsberg of the U. S. National Museum described it as "like no other fish scale I have ever seen." He believed it came from a coelacanth or a similarly primitive, yet-unclassified relative. Unfortunately, Dr.

Ginsberg never heard from the shop owner again, and the fisherman was never traced.

Zoologist J. Richard Greenwell has unearthed reports of similar scales being collected from a "trash fish" pile from a Gulf shrimper around 1973 and being found in a Biloxi, Mississippi souvenir shop in 1992. In neither of these cases was a scale examined by a proper authority. Still, even if Dr. Ginsberg's scale represents the only verified evidence, it did exist...and it had to come from somewhere.

Traveling Ants

It's one thing to discover a new species in some remote jungle or deserted island. It's quite another to have it come to you and literally walk onto your desk. But that's what happened to Kathryn Fuller, president of the American branch of the World Wildlife Fund.

In 1989, she noticed small yellowish ants crawling on her desk in her Washington, D.C. office. She paid them little attention at first, and they were still appearing in 1990 when she pointed them out to visiting Harvard entomologist Edward O. Wilson. It was a good thing for science Ms. Fuller hadn't called an exterminator, because the ants were of a species never seen before. They had apparently entered the U.S. from the Carribean along with the potted plant they inhabited.

So it was that the species now known as *Pheidole fullerae* entered the records of entomology.

The location of this new find may have been surprising, but the actual discovery of a new insect is almost an everyday occurrence. According to the eminent Dr. Wilson, author of the recent *The Diversity of Life,* insects and other small invertebrates form the vast majority of the 1,032,000 known species of animals. They also form the bulk of the undiscovered species, which may number as high as 30,000,000.

Accordingly, it's not terribly difficult to discover a new species.

Dr. Wilson reports that, in the rain forests of the Americas, "The chances are high, in fact certain, of finding a new species or phenomenon within days, or, if you work hard, hours after arrival."

Entomologist Terry Erwin has been proving the truth of Wilson's statement on a grand scale. By fogging tropical trees and collecting the insects that fall from the rainforest canopy, Erwin has collected more than 5 million insects and other creatures in the last twenty years. Incredibly, most of these represent new species. Erwin's finds have piled up so fast that he has no time to assign Latin names to new species, so he uses a code of letters and numbers and leaves the formal naming to some future scientist in less of a hurry to survey threatened habitats.

Discoveries of invertebrates generally don't get much attention, but some of them are certainly noteworthy.

In May 1961, Mrs. Marie Lantham, an American explorer and animal collector, literally dug up an invertebrate that did get some press. Her find was a new type of giant earthworm—a new genus, in fact—living at an altitude of 14,000 feet in the Columbian Andes. "Gertrude" (also known, confusingly, as "Willie"), a live specimen over five feet long, was presented to the London Zoo.

Then there's the latest addition to the bee family. Everyone knows about the vital role bees play in the pollination of plant life. However, there is one very strange insect, the vulture bee, which has eschewed that noble task entirely.

This unique species, described from Panama in 1982, shows no interest in pollen. Instead, it has mandibles equipped with large teeth to attack the carcasses of dead animals of all sizes. David Roubik, who discovered the insects, experimented by placing the remains of his Thanksgiving turkey near a nest. Hundreds of the vulture bees swarmed over the carcass, and in a matter of hours there were only bones remaining.

The carrion-eater discharges an enzyme which aids in decomposition, allowing the bee to chew and consume flesh. Back at the hive, the regurgitated material is passed to workers who process it into food. Interestingly, despite the fate of Roubik's turkey, the bees show an apparent preference for amphibian carcasses over other types of carrion.

It turns out the species ranges through low-lying rainforests from

Panama down to the Amazonian region of Brazil. The stingless bees' place in the ecosystem is an interesting one. Roubik notes the switch to animal food is an advantage in a range where frequent rains impede pollen-gathering, and the insects have found a niche where no other bees compete with them.

Finally, a treasure trove of insects and other invertebrates was found in 1986 when the first explorer descended into a Romanian cave. Isolated from the surface for millions of years, the cave developed an ecosystem like no other on Earth, where the air was only seven to ten percent oxygen and chemosynthetic bacteria formed the basis for the ecosystem. At least twenty-four unique new arthropods adapted to these conditions have been collected so far, including a blind water scorpion that breathes through its tail.

Itundu, the Congo Peacock

A scientific achievement might be called "a feather in one's cap." In the unique case of the Congo peacock, that phrase is a literal description of its discovery.

In 1915, Dr. James Chapin of the American Museum of Natural History returned from an African expedition with a large collection of specimens. One was a feathered native headdress. When Chapin found time for a careful examination of these feathers, he found that one could not be identified.

It was still a mystery in 1936, when Chapin visited the Congo Museum in Belgium. There he spotted two stuffed birds with black-barred wings sporting the same strange feathers. The creatures were labeled as young peacocks, but Chapin realized this was an error. He learned that the birds had been acquired from the Belgian Congo (now Zaire) in 1914, but no one had paid them much attention.

Somebody should have, because the birds belonged to a new species. The excited Dr. Chapin returned to Africa in 1937. There a native hunter took him to look for the bird the locals knew as *itundu*.

On July 16, Chapin became the first western scientist to see a living Congo peacock.

It was a sight well worth the trip, for the Congo peacock (which technically is a pheasant) is a spectacular bird. The male is decked out in metallic blue, green, and black feathers, topping off this outfit with a unique tuft of white bristles on its head. The female, attired in flashy orange-red, striped vertically with green, is no fashion slouch herself.

This was hardly Africa's last avian surprise. A bird collected while sleeping in Zaire's Itombwe Mountains in 1951 was described as a new species, the Congo bay owl (or Itombwe owl). This African relative of the common barn owl is a nocturnal forest resident, and may be even rarer than the Congo peacock. Since Chapin's encounter, only a few people have seen the peacock in the wild: aside from an unconfirmed sighting from Burundi over twenty years ago, it appears that no one has seen the owl.

New Parrots of South America

Any reference book more than a few years old will give 1914 as the date the last new parrot in the Western Hemisphere was found. Today, those books need some revisions.

One reason is a discovery made in 1988, when ornithologist Robert Ridgely of The Academy of Natural Sciences in Philadelphia was studying the birds of Ecuador. In a cloud forest on the western slope of the Andes, he spotted a small, previously unknown green parrot with a crown of red feathers.

Scientifically described as *Pyrrhura orcesi,* the bird apparently lives only in the province of El Oro at altitudes between 1300 and 2600 feet. The El Oro parrot, as Ridgely named his find, remains little-known and appears to be rare. The Ecuadorean government has extended protection to the bird, but it is still threatened by habitat destruction.

Surprisingly, the long drought of parrot discoveries has ended not with one new bird, but three. After Ridgely's find, Brazilian scientists reported a new species from the Amazon lowlands in 1989.

Dr. John O'Neill had actually scooped both these discoveries by finding a new parrot in Peru in 1987, although his bird wasn't formally described until 1991. O'Neill is something of a legend in ornithological circles, having described no less than eleven new Peruvian birds in his career. These include two owls and the splashy orange-throated tanager, which O'Neill discovered in 1963 when he was only a college student.

Writer Don Stap chronicles the 1987 find in his fascinating book *A Parrot Without a Name*. As Stap tells the story, O'Neill's expedition was nearing its end. His team was preparing specimens for shipment home when one member shot two small green parrots and handed one to O'Neill with the standard question, "What is it?"

O'Neill turned the bird over in his hands for a minute. "It's nothing," he said calmly. "It's something new." It was indeed something new, a parrotlet with a distinctive blue forehead, and its discovery marked O'Neill's twelfth addition to Peruvian bird lore. (A "parrotlet" is basically a small parrot, although the definition is a bit fuzzy.)

"You shall know a bird by his cry...and all birds have their voices," wrote D. H. Lawrence, and so it was that ornithologist Ted Parker found another new bird in the fruitful Peruvian forest. Parker spent years recording the calls of thousands of bird species throughout the Americas. In 1982, in northern Peru, his highly trained ear picked up a bird song Parker knew he'd never heard. He followed that song to his discovery, the orange-eyed flycatcher. It was the first New World flycatcher described in a hundred years. Tragically, Ted Parker died in a South American plane crash in August, 1993, while pursuing the work he loved.

Dr. Ridgely estimates that, in recent years, two to six new birds have been described annually from South America alone. If we can save the tropical forests from destruction, there is every reason to expect more exciting finds like these.

Archey's Frog

Among New Zealand's seemingly inexhaustible supply of bizarre animals, a tiny frog might well be overlooked. And the amphibian now known as Archey's frog was indeed overlooked for a long time. It was finally described only in 1942, although sightings had apparently been recorded as much as eighty years earlier by observers who didn't realize the uniqueness of what they were seeing.

Archey's frog is a little beast, rarely over an inch long, with green and brown skin and golden eyes. It lives only in forests on the island of Coromandel.

Once found, the frog proved to be a very interesting specimen. For one thing, it lives mainly on land: in fact, it has adapted so completely to a terrestrial environment that there is no webbing between its toes. Even more unusual are its reproductive habits. Archey's frog skips the tadpole stage entirely. Eggs are laid under rocks, where the male frog guards them until the little ones emerge as fully developed miniature frogs.

Biologist Edward Kanze writes that Archey's frog, along with its closest relatives, is an important proof of continental drift. It is a member of a group known as the "tail-wagging" frogs, which are actually tailless like all other adult frogs but retain the muscles used for tail-wagging. There are two other tail-wagging frogs in New Zealand, along with one in North America. Frogs, even the swimming varieties, cannot cross oceans, so the distribution thousands of miles apart of these four related amphibians indicates they once all lived on a single land mass that split up. Archey's frog is a reminder that even an apparently insignificant animal may have a lot to teach us.

The anurans (frogs and toads) have produced other surprises since Archey's frog was found. In 1942, the first of three species of live-bearing toads was found in Guinea. This amphibian is confined to a few square miles of rocky ground. There it spends nine months of the year hiding in holes and crevices, emerging only during the wet season.

Nine years later, in Columbia, one of the world's largest toads was discovered. The Columbian giant toad boasts a head-and-body length up to ten inches, making it the longest (although not the heaviest) toad known to science. This striking amphibian, decorated with a copper-colored back and brownish underside, is now considered endangered.

In 1962, two researchers discovered the Andes toad in a most peculiar environment. While hiking through Chilean snowfields a mile above sea level, they found the toads happily spawning in sulfur springs with a water temperature of 90 degrees Fahrenheit.

A tiny Ethiopian species discovered in 1978 has adapted to equally strange surroundings. Malcolm's earless toad lives in the bitter cold and frequent drought of the Bale Plateau at altitudes above 9500 feet. Due to the harsh conditions, the species has adopted the non-amphibian habit of fertilizing its eggs internally. The eggs are then deposited in soil, where the tadpoles live off a yolk sac until they metamorphose into adults. These are remarkable adaptations, and how they came to be is a puzzle.

A frog with an even stranger characteristic turned up in Australia about ten years ago. It is the second known species of gastric brooding frog, which means just what it sounds like: the eggs are incubated in the stomach. Ironically, the new frog was discovered in a search for specimens of the *first* known gastric brooding frog, which is apparently extinct. The new frog will take the old one's place in studies by physicians, who are fascinated by its ability to switch off its digestive secretions.

Finally, we have *Rana subaquavocalis,* a very rare frog whose Latin name describes it well. Biologist James Platz collected the first specimen of this type of leopard frog in Arizona's Huachuca Mountains in 1990.

It is the only known species of frog to issue its mating call underwater. This seems to be an advantage for the frog in that its potential mates only need to listen underwater, where no other species' mating calls will clog up communications. This also prevents predators (and herpetologists) from homing in on the frog's vocalizations, one reason this interesting amphibian eluded science until the present decade.

The Kouprey

In 1936, Achille Urbain, the director of France's Vincenneso zoo, was visiting Cambodia when he saw what appeared to be an oversized set of ox horns in the possession of a local veterinarian. Learning the trophy came from a wild bovine known locally as the kouprey, Urbain asked his colleague to secure a live specimen.

The vet, Dr. Sauvel, was able to fulfill the request the very next year. When the large, dark grayish animal emerged from its shipping crate in France, Urbain and other experts were as puzzled by the kouprey as the kouprey undoubtedly was by the Frenchmen. Six feet high at the shoulder, with a huge dewlap, the animal engendered speculation that it was some kind of hybrid arising from domestic stock. Urbain finally satisfied himself that it was a genuine species, and he authored a paper formally introducing the beast to Western science.

The kouprey was the largest land animal discovered in over three decades, and no larger ones have been confirmed since. A big bull may weigh almost a ton. Females have lyre-shaped horns, while the bulls' horns spread more widely and have frayed tips: the animals reportedly have a puzzling habit of rubbing them against termite mounds.

Not much else is known about the kouprey. The French specimen remains the only one ever held in captivity in the West. Even Urbain's identification of the animal has been challenged: although the kouprey's status as a species is generally accepted, a minority of experts believed it did have domestic ancestry, while an opposite opinion held that it was so unlike anything else it belonged in its own genus.

The kouprey was obviously rare, and in the ensuing decades it became a lot rarer. By 1969, the wild population was estimated at 100 or less. The animal had the misfortune to occupy a range made up of Laos, Vietnam, Thailand, and Kampuchea (Cambodia). It lived in small herds and preferred fairly open country over the dense forest that might have provided better concealment.

As wars continued to sweep over the region, the animals were shot to feed soldiers, guerrillas, or the desperate peasants who had been deprived of their own food by the soldiers and guerrillas. Others were unintentional victims of artillery barrages or land mines. Such depredation led to fears that the kouprey was extinct.

Fortunately, postwar sightings beginning in 1982 have confirmed the species is hanging on. The 1982 rediscovery involved a group of five in northeast Thailand. In Kampuchea, sightings were reported in 1984 by villagers who are known to value the animal's genes: they leave their domestic cows in the forest, hoping they will mate with kouprey bulls. Recent sightings of a few animals have also been confirmed in Laos and Vietnam.

Today the Species Survival Commission of the World Conservation Union estimates there are between one hundred and three hundred kouprey scattered across the four countries. A Kouprey Action Plan has been drafted to try to preserve the species, which is now considered the rarest large mammal in the world.

The Chacoan Peccary

The largest land-dwelling "living fossil" to be discovered in the last two decades is the Chacoan peccary. Until recently, this wild pig was known only from South American fossil remains. As far as science knew, it had passed from the scene thousands of years ago.

Natives of the Gran Chaco area of Paraguay, Argentina, and Bolivia always knew there was a type of pig roaming their sparsely inhabited neighborhood, but had no idea the zoologists of the world might think the animal important. To them, it was just another (edible) part of the local fauna.

The peccary remained unknown to science until a National Geographic Society expedition led by Dr. Ralph Wetzel penetrated the Paraguayan Gran Chaco in 1975. They discovered the peccary was not only alive but doing quite well. The scientists returned with twenty-nine specimens, and Dr. Wetzel described the population as "significant."

The Chacoan peccary is larger than its known living relatives, the javalina and the white-lipped peccary. A large Chacoan peccary may be almost four feet long and weigh a hundred pounds. The animal is brownish-grey except for a pale collar-like band around the shoulders. The head features a long snout armed with fierce-looking tusks. This extended snout makes room for oversized nasal passages, an adaptation to the arid climate of its habitat. To make a living in this desolate area, the pigs eat any type of vegetable matter available, including cacti.

The peccary has always been hunted, but the major threat to its survival comes from development. Road-builders and cattle ranchers are braving the inhospitable region, and conservation measures are needed to make certain this one-time "fossil" doesn't revert to extinct status permanently.

The Vaquita

The story of the world's smallest and rarest porpoise begins with a single skull found on the beach in the Gulf of California. That discovery was made in 1950, but another eight years passed before Kenneth Norris and William McFarland had enough information to present the vaquita, or Gulf of California porpoise, to the scientific world.

At five feet long or less, and never weighing much over a hundred pounds, the vaquita is tiny by cetacean standards. Its size may have helped it keep hidden: so, undoubtedly, did its shyness. The animal does its best to avoid boats, an unusual trait for a porpoise. Local fishermen did know it existed, and it was they who called it *vaquita*, or "little cow."

This porpoise is mainly light gray, although the color usually darkens from the dorsal fin to the tail. The belly is white, and there are dark ovals around the eyes.

The vaquita is rare and undoubtedly endangered. Many animals have been killed accidentally in gill nets. The Gulf's ecology has suffered due to overfishing and agricultural runoff, and the surviving porpoises' food supply is dwindling. The vaquita is unusually vulnerable to such threats because it does not migrate: in fact, it has the most restricted range of any marine mammal in the world.

Estimates of the remaining population range from 200 to 400 individuals. Since the bashful porpoises are hard to count accurately, there may be even fewer than 200 left. It's another unfortunate example of a species which may become extinct before science can even document its decline.

New Worlds Beneath the Sea

Finding an unknown animal is a rare and exciting event in a scientist's life. Finding an entirely new environment filled with undescribed species is a thrill few researchers even dare to dream of.

Dr. Robert Ballard was leading an expedition off the Galapagos Islands in 1977 to study seafloor spreading, part of the process of continental drift, when his team found something completely unexpected. An instrumented sled recorded a large patch of clams and mussels, some a foot wide, living on the sea floor in an area that also produced peculiarly high temperature readings.

The submersible *Alvin* was brought in. Almost nine thousand feet down, its crew stared at a landscape from a science-fiction film. Geothermal vents gushed hot, mineral-laden blue water from the ocean floor. Surrounding the vents was a densely packed colony of animals, many of which existed nowhere else. Tube worms, looking like eight-foot-long white hoses with red feather-dusters poking from their open ends, clustered at the site. Crowded around them were equally strange creatures such as white clams, yellow mussels, and ghostly-pale crabs. Pink fish with blue eyes nosed around the fringes of this outpost of life.

The tubeworms received the most attention. When the first ones were returned from the Galapagos vents, biologists were thoroughly puzzled. The animals had no mouth or digestive tract, just a long section of brown, spongy tissue. The question of how the things fed stumped the world's best scientific talent until Colleen Cavanaugh, then a "mere" graduate student, theorized the worms were hosts for symbiotic bacteria which utilize the chemical-laden water to produce nutrients for themselves and the tube worms. The white clams found at the vent sites have a similar arrangement.

In fact, the entire vent colony is based on bacteria. These adaptable little chemical factories form thick mats on the seafloor in addition to residing in worms and clams. The community's higher animals live on the bacteria themselves, on bacterial byproducts, or on other animals living on the bacteria. This is the first ecosystem ever found that does not in any way depend on photosynthesis, the sun-powered "engine" which drives all other life on earth.

The tube worms are so unusual they've been assigned their own phylum, Vestimentifera. The entire animal kingdom includes only some thirty phyla. To add a new one, as Dr. Meredith Jones did for the tube worms, is an event assuring scientific immortality. (It should be noted that some experts don't agree with this, instead making Vestimentifera an order within the phylum Pogonophora.)

Since then, other vent communities have turned up, providing scientists with a continuing flow of previously unknown animals. 300 new species, requiring the creation of ninety new genera and twenty new families, had been collected by 1994. These included some finds nearly as bizarre as the tube worms. Vents streaming smoke black 650°F water on the Mid-Atlantic Ridge are home to a kind of eyeless shrimp equipped with a light-sensing patch on its back. On the vertebrate side, oceanographers studying a vent off

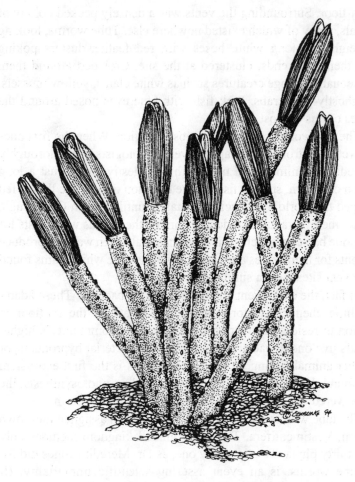

Baja California collected weird pinkish white eel-like fish about a foot long.

It turned out that hot-water vents were not the only features which sustained chemosynthesis-based communities. A site named Bush Hill, in the Gulf of Mexico, is clustered around a petroleum seep. Another community in the Gulf consists of a ring of mussels surrounding a brine pool, a seep which leaks water more than three times as salty as seawater.

In 1990, the first fresh-water hydrothermal environment was found. Its vent is located at a depth of a quarter-mile in Russia's enormous Lake Baikal. Like the oceanic vents, this one supports a variety of fish, sponges, and worms, including some new species.

These unearthly habitats have opened up new horizons in the study of geology and biochemistry as well as zoology. The effort to understand such colonies and classify their inhabitants will continue for many years to come.

The Amazing Animals of Vu Quang

The idea of a "lost world" was first popularized by A. Conan Doyle's novel of that name. Practically everyone, though, has a romantic fondness for the idea of a realm untouched by humanity and time, a place where new and strange animals roam.

Such regions do exist. Vietnam's Vu Quang Nature Reserve, for instance, may not measure up to the dinosaur-infested jungle of Doyle's imagination, but scientists who first surveyed it in 1992 found enough to startle them.

In this undeveloped (and unbombed) area along the Laotian border, an international team catalogued a wealth of animals, including many rare species. They also found four new animals. These included a fish, a sunbird, and a handsome yellow box turtle. The fourth discovery was the mammal originally called the Vu Quang oryx—easily the zoological find of the year. •

The search for this new animal began when scientists visited some of the area's few indigenous inhabitants and saw strange horns decorating the walls of hunters' homes. When questioned, the hunters described an animal known by a variety of names which included "forest goat," "mountain goat," or, most commonly, "*sao la*" (spindle horn). They reported the animal was mainly a leaf-eater, and was usually found in mini-herds of two or three animals. The seeming contradiction between the "mountain" and "forest" names arose because the beast is a seasonal migrator, living in the hills in summer and the lowlands in winter.

The creature's true identity was not immediately determined. All the investigators initially knew was that their quarry was a hoofed animal whose horned skull was about three feet long. The horns were long, straight, and almost parallel. Dr. John MacKinnon, the British leader of the World Wide Fund for Nature (WWF) team, commented, "There are only two animals in Asia vaguely like it, and nothing with parallel horns like these."

The team brought back three skulls, along with skin samples for DNA testing. Much to the researchers' disappointment, they failed to observe the live animal.

Possible relatives MacKinnon speculated on included the serow, a type of Asian goat, and the anoa, an Indonesian dwarf buffalo. When the DNA evidence was in, these guesses had to be discarded. MacKinnon had placed the creature in the right family—the bovids, which include cattle, antelopes, and goats—but it was an entirely new genus and species, apparently belonging in the same subfamily as oxen. MacKinnon called his discovery "a primitive cow that's had a fairly undisturbed evolution." He and his Vietnamese colleagues erected the genus *Pseudoryx* for the species and gave it the informal name of Vu Quang oryx.

Vietnamese scientists have now collected partial specimens from more than twenty animals, and a taxidermist in Hanoi has created a representation of the whole creature. An adult *Pseudoryx nghetinhensis* is about three feet high at the shoulder and weighs over 200 pounds. The females have horns almost perfectly parallel to each other, while the longer horns of males diverge at a slight angle. The animal is reddish brown or dark brown, with a variety of white stripes and patches on the face.

In June 1994, the first live specimen was obtained when Vietnamese scientists captured a five-month-old calf. A second calf was captured a few months later, but both animals died that same year. About this time, a change in the animal's English name was taking place, with authorities now calling it the "Vu Quang ox." This is taxonomically more accurate, although the animal *looks* more like an oryx than it does anything else.

Estimates of the living population place it as high as several hundred. This is still a dangerously small number for a mammal, and this exciting discovery presents conservationists with a test case.

Can Vietnam and the international conservation community work together to control development, poaching, and other threats, and preserve such a find in its natural habitat?

The discovery of a large land mammal is perhaps the most publicized event in zoology. The last such find before the forest goat was a new deer, a member of an Asian genus called the muntjacs. Muntjacs, which are generally small and have short, thick antlers, are known as "barking deer" because of the deep barklike sound they emit if threatened. The new species was reported by Chinese zoologists Shi Liming and Wang Yingxiang from the mountains of China's Yunnan province and adjacent Tibet in 1988. Two mammologists had already described another new muntjac from Borneo in 1982.

In 1994, still another species of muntjac was found. This discov-

ery also began with unidentifiable antlers obtained from the Vu Quang region. Dr. MacKinnon was nearing the end of a return expedition to Vietnam in March of 1994 when local inhabitants showed him and researcher Shanthini Dawson some skulls from another animal they hunted. MacKinnon recognized the skulls belonged to some type of muntjac, but the species was a new one. The structure of the horns was unique, and the skull was much larger than that of any known muntjac. MacKinnon estimated the new "giant muntjac" weighed more than 100 pounds, twice the size of the common muntjac.

Also in 1994, researchers working for the Wildlife Conservation Society (WCS) became the first Westerners to actually see the giant muntjac. They found a specimen in Laos, in a private menagerie maintained by a local military commander, and were allowed to take a blood sample for DNA analysis.

Dr. MacKinnon believed the red-furred giant muntjac deserved its own genus, and he named it *Megamuntiacus vuquangensis*. Alan Rabinowitz of the WCS, however, reported in August 1994 that the blood analysis done so far indicated the new species was close enough to other muntjacs to belong with them in the genus *Muntiacus*, so this point is not yet settled. Also reported from the private Laotian zoo was a hybrid of the giant and a more common species of muntjac, which would support the same-genus line of reasoning.

That wasn't the end of the new animals from southeast Asia. From a mountainous area north of Vu Quang, called Pu Mat, have come skulls of an animal with short, sharp antlers. This is known locally as the "slow-running deer" or "slow-moving deer." Another set of antlers collected by MacKinnon apparently belongs to an animal known as the *mangden*, or "black deer," although no additional evidence has been gathered as of this writing.

Unbelievably, still more surprises were in store. When, in 1995, Western zoologists revisted the captive giant muntjac, they found in the same zoo still *another* muntjac, a very small, almost black animal that also appears to be a new species. Back across the border in Vietnam, much farther south, a set of horns for sale in Ho Chi Minh City led to the 1995 description of one more new animal, known locally as the *linh duong*, or "mountain goat." It is still known to science only by its horns, which resemble high-rise motorcycle handlebars.

Finally, two other sizable land mammals have been presented to science in recent decades. The bilkis or Yemen gazelle was first described in 1985. Its case is a reminder that "described" doesn't always mean exactly the same thing as "discovered." The specimens involved—skins and skulls from a total of six animals—had actually been collected from Yemen in 1869 and 1951. They rested in Chicago's Field Museum of Natural History, waiting for scientists with the time and interest to examine them closely.

Eventually, mammologists Colin Groves and Douglas Lay did

just that. They distinguished the new species by its dark, solid coloration and upright horns, among other peculiarities. Soon afterward, ten living members of the species were identified in a privately owned herd in Qatar.

The red goral, a brownish red wild goat with a darker stripe running down its back, was recognized in 1961. Once again its existence was deduced from old specimens, one having been made into a rug. The animal's range includes—or included—Tibet, Assam, and northern Burma, but the exact boundaries are uncertain.

In the wake of the Vu Quang finds, a British newspaper quoted Dr. MacKinnon as saying these discoveries are evidence that similarly unexplored regions around the world could yield still more large animals. Let's hope so.

Neopilina: the Strange Little Fossil

Everyone is fascinated by the idea of "living fossils," animals which have survived unchanged while the evolving world passed them by. The coelacanth is the most famous: others include the tuatara, a lizardlike New Zealand reptile whose three eyes (the third is degenerate but functional) watched the dinosaurs come and go.

The term "living fossil" is not reserved for vertebrates. Among the myriad specimens dredged up by the famous *Galathea* expedition in the early 1950s were ten limpetlike shelled animals. They came from sea-bottom mud 11,878 feet beneath the surface off the Pacific coast of Costa Rica.

What were they? No one was sure. The new discoveries had pale yellow shells with an oval shape, about one and a half inches long and half an inch high. A large foot (colored pink and blue) was surrounded by five pairs of primitive gills.

While the shell and teeth said "mollusk," the gill regions showed a segmented construction resembling annelid worms. Was the new animal either of these, or was it something entirely unique? It most resembled a model that biologist Brooks Knight had created showing what the ancestor of today's mollusks might have looked like. But that hypothetical animal—no actual fossil had ever been found—was presumed to have died out 350 million years ago.

Neopilina galathea filled an important gap in the evolutionary record. When its identity was finally sorted out, the little critter was literally placed in a class by itself. Since then, three more *Neopilina* species, each represented by a single specimen, have been dredged from the depths. The scientific detective work of finding more examples and determining their exact place in the parade of evolution goes on.

A 1955 find from shallow water provided another missing piece in the evolution of our planet's incredibly complex fauna. Biologist Howard Sanders' discovery, *Cephalocardia,* was a miniature shrimplike creature apparently dating to the beginning of crustacean history, over half a billion years ago.

A pioneering exploration vessel we have already mentioned, the submersible *Alvin,* pulled in a surprise in 1979. Near a hydrothermal vent in the eastern Pacific, researchers on the sub collected a strange-looking stalked barnacle, the stalk serving to allow these normally fixed creatures some degree of mobility. It had never been seen before, even as a fossil, but apparently belonged to a group which flourished before the dawn of the age of reptiles.

The waters off New Zealand produced a similar surprise in 1985. Clinging to sunken logs over three thousand feet below the surface was a round animal less than half an inch wide. Named the sea daisy, it appeared to be a distant relative of the starfish, even though only traces of the classic five-pointed starfish design were apparent. That was enough to put it into the same phylum, the echinoderms, but it proved very difficult to classify this diminutive invertebrate more precisely.

The sea daisy is spiny on top, and its underside is covered by a flat membrane that biologist Michael Bright compares to plastic wrap stretched over an upside-down saucer. The sea daisy, too, was assigned its own class. There was just nothing like it, except for fossils predating the dinosaurs.

Fossils from the same period included the graptolites, tiny colonial creatures who built homes of collagen secretions layered in strips like mummy bandages. At one end of each inch-long communal house, a peculiar sharp spike rose like a TV antenna. Graptolites were presumed to be related to modern homebuilders called pterobranches, but pterobranch dwellings lacked the characteristic spike.

After an apparent absence of 300 million years, graptolites resurfaced. In 1992, French researchers sent a sampling of seafloor specimens to Dr. Noel Dilly, a London ophthalmologist whose "hobby" of studying pterobranches grew on him until he became one of the leading experts on the animals. Dilly's first reaction was, "Not another boring collection to hack through." His second was, "I don't believe this." He was looking at characteristic graptolite dwellings, spikes and all.

How could animals encased in such structures build the still-puzzling spikes? Dilly found that graptolites have a characteristic that sets them apart from other home-building seadwellers: they can crawl outside and add onto their homes, scrambling over the "roofs" like miniature suburbanites putting on new shingles. The spike is apparently built up as the filter-feeding animals take turns using it as a vantage point to pluck food from the seafloor currents.

The graptolite is a reminder that not all animals evolve: some just find a comfortable ecological niche and settle down for a long stay.

The remipedes have adopted the same philosophy. Remipedes are curious creatures an inch long that are, in the words of biologist Jill Yager, "at the base of the crustacean family tree." Yager discovered these survivors in the Bahamas in 1979 and later on Mexico's Yucatan peninsula. They congregate in water-filled caves, living just below the dividing line between the fresh water draining down from the surface and the salt water seeping in below. Yager notes that such caves are, from a biologist's viewpoint, still a little-known realm.

Recent finds concerning minute yet significant animals include two which may not qualify for the "living fossil" label, but are so unique that new phyla had to be erected to house them. A Russian expedition in the early 1960s dredged up countless specimens of wormlike sediment-dwellers which were labeled pogonophores. Similar creatures had actually been found forty years earlier by the research ship *Discovery II,* but those specimens were dumped overboard as annoying, net-clogging "fibers." Pogonophores are even strange on the inside: they have no digestive tract of any kind, but appear to absorb nutrients directly in the form of dissolved organic matter found in the water and sediment of their environment.

Another phylum was established in 1983 for an animal a mere 1/100 of an inch long. Called *Nanaloricus mysticus,* this find resembles a legless miniature hedgehog. Burrowing through the ocean

46

floor sand at depths of thirty to 1,500 feet, it probes with its prominent snout for even smaller prey animals.

Danish zoologist Reinhardt Kristensen proposed the phylum Loricifera for this totally new organism. That phylum's population now includes over two dozen species. Kristensen named one of these for American scientist Robert Higgins, who had predicted the existence of such animals. Higgins pronounced himself "very pleased...even though it is such an ugly creature."

How many more finds are waiting for us in the ocean depths? Well, Frederick Grassle of Rutgers University led an effort to collect over two hundred core samples of the Atlantic seafloor in the mid-1980s. When all the sediment had been sifted, the somewhat flabbergasted scientists found they had collected *460* new invertebrates.

We shouldn't forget the mollusks, either: about 600 species are discovered each year, either in the field or during reexaminations of old collections.

The tiniest animals play as important a role in our ecosystems as the largest, and discoveries like this give us some idea of how complex the web of life really is.

The Tamarin and the Marmoset

Brazil houses the world's largest tropical rainforest, and no one is surprised to learn there are still new species to be found there. But no zoologist would have predicted we'd find three new monkeys in the last decade alone.

The recent finds started with a new squirrel monkey, a two-pound creature presented in 1985 with the scientific label *Saimiri vanzolinii*. This primate inhabits a tiny territory, perhaps the smallest of any South American monkey, at the joining of the Amazon and Japura rivers in central Brazil.

The next discovery was made far from the squirrel monkey's haunts and involved an even smaller creature known as the black-faced lion tamarin. This monkey, not much larger than a rat, was found on a coastal island named Superagui by two Brazilian scientists engaged in cataloging the local bird life. A fisherman described the black and gold monkey to the visitors, professors Maria Lucia Lorini and Vanessa Guerra Persson, and provided a skin. Lorini and Persson then found the live animals and, after a month of observing them in the wild, published their results in June 1990.

This colorful animal is also known today as the caissara monkey. The local fishermen, or caissara, had known about it for a long time, and the official discoverers acknowledged their debt by proposing *Leontopithecus caissara* as the species' formal name.

The location was a surprise: Superagui is largely developed, and lies in the heavily populated region surrounding Sao Paulo. Primatologist Russell Mittermeier, president of Conservation International, marveled that it was "almost like finding a major new species in the Los Angeles suburbs."

The discovery brought the number of known species of lion tamarins to four. All are endangered: the golden-rumped species was rediscovered in 1970 after no one had seen it in sixty-five years.

The most recent find came from the depths of the rain forest. In the fall of 1992, the first report of the Rio Maues marmoset was published. The new marmoset is a long-tailed grayish furball weighing less than a pound. It sports faint black stripes, large eyes, distinct ear tufts, and a cuddly face reminiscent of a koala.

Swiss scientist Marco Schwarz actually found the monkey in 1985 near the Maues river, a tributary of the Amazon. He obtained a live pair of the marmosets, but didn't immediately recognize their uniqueness. It wasn't until photographs of the eight-inch-long animals reached Dr. Mittermeier that the species *Callithix mausei* was officially added to the primate order.

Brazil's known primate count now stands at an impressive sixty-eight species. Dr. Mittermeier thinks four or five more will be found before the turn of the century. That sounds startlingly optimistic, but Mittermeier knows something about finding primates, having personally rediscovered the supposedly extinct Peruvian yellow-tailed woolly monkey in 1974.

Other new primates have popped up in recent years, including two lemurs from Madagascar. The golden bamboo lemur was first sighted in 1986, when a Duke University team searching for a rare known species happened across it. Three years later, the golden-crowned sifaka was classified, concluding a scientific detective story that began with British primatologist Ian Tattersall's 1974 photograph of a mostly-white lemur he couldn't identify.

Going back a few decades, a striking primate discovery was announced to the world in 1953. In that year, naturalist E.P. Gee published the first description of the golden langur from Assam. The langur's coat shines a pale gold, except for a dark mask on the animal's face.

Little is known of this slender-bodied, long-tailed species even today. Its obscurity is surprising, since it's a large monkey (up to two feet long, plus the tail) which is active mainly in daylight. It is, like most langurs, a tree-dweller that spends most of its time munching on the vegetation. The golden langur's range is limited to small areas of India and Bhutan.

A year before the golden langur made its scientific debut, its cousin, the white-headed langur, turned up in China's Guangxi province. Both primates are now rare and endangered. More recently,

China produced a large race of the Assam macaque which may qualify as a separate species. This was identified from a captured specimen whose finders originally thought they'd nabbed a *yeren,* China's presumably mythical "wildman."

Africa has plenty of monkeys of its own, including new ones. The Salongo monkey was discovered in Zaire in 1977, when a Japanese tourist bought a skin from native hunters. The grayish brown primate proved to be a new species belonging to a wide-ranging and prolific genus called the guenons. A new subspecies of the redtailed guenon was discovered in 1984 in a very different location—a French pet shop. Another new species, the sun-tailed guenon, was described in 1988 after being found as a captive in a village in Gabon. British primatologist Mike Harrison was the first to see this monkey, known in its very limited range as the *mbaya.*

This wasn't the only case where a monkey was well known to the local inhabitants before scientists "discovered" it. In 1985, zoologists Katherine Homewood and W. Alan Rodgers spotted a new gray-brown monkey near Sanje Falls in Tanzania. Their native guide was puzzled by their excitement. It was just a *ngolaga,* he said: if they wanted to see one, he could just have shown them the tame one in his village and saved them all the arduous hiking through the forest. They found and bought this pet, and thus the Sanje mangabey joined the known primates.

Given this track record, anyone writing a textbook on primates of the world would be wise to leave the last pages blank.

Giant Jellyfish of the Deep

The more we "do business in the great waters," as the Bible puts it, the more we are reminded that much of the life in those waters is still to be discovered. Recent finds include some noteworthy additions to our inventory of marine invertebrates.

The tranquil blue expanse of the Eastern Pacific yielded a major surprise in 1989. Divers photographing marine life for an episode of *Nature* found themselves swimming among purple jellyfish of a type never seen before.

The jellyfish had not remained unknown due to small size. They were three feet across, and their arms trailed for twenty feet. Nor were they extremely rare: the divers saw hundreds of them, and others later drifted up on the beaches of Baja California. Apparently, they had simply kept to themselves, well out in the ocean, until a change of currents brought them into the light of science.

A typical specimen of this new species, placed in the genus *Chrysaora,* displays a variety of purple, pink, and whitish hues. Long, slender tentacles trail from the edges of the pulsating bell, and in the center are the more substantial "oral arms," pleated, scallop-edged appendages that sting and trap prey to provide sustenance.

As the jellyfish drifts through the tropical seas, it becomes the center of a mobile community. Shrimp, crabs, and other small creatures ride on the arms or drift among them, somehow unconsumed by their host but deriving protection against predators that are not so immune. Some even steal their host's food. Other fish are equipped to nibble the jellyfish itself, so the creature, like most predators, is also prey.

Interestingly, this is not the only modern discovery of a new species of three-foot-wide jellyfish. Another had turned up three decades earlier in the Indian Ocean.

Closer to home, investigations of a deep canyon in California's Monterey Bay have yielded a string of new invertebrates. These include a squid that is completely transparent and a siphonophore (a colonial gelatinous creature) 120 feet long.

On the other side of the Americas, Bermudan fisherman John Ingham recently introduced the world to a new crab from his native waters. About ten years ago, he experimented with setting crab traps at much greater depths than usual—over four thousand feet down. He pulled up full catches of red deep-sea crabs, the largest of which weighed sixteen pounds. The new species, now known as *Geryon inghami,* is related to a Florida type known as the golden crab. In 1985, another golden crab relative was identified—a new species named *Geryon gordonae,* from the opposite side of the Atlantic.

Some new octopi were also spotted for the first time in the

1980s. These include two new Atlantic species of the gelatinous deep-sea octopus *Allapossus,* a type of short-armed, large-eyed octopus known to have a total span of up to seven feet. Gelatinous octopi are so called because their flesh is soft and translucent, as if they were jellyfish pressed into an octopus mold. One of the new finds, larger than any previously known gelatinous species, sports eyes six inches across.

In 1984, French oceanographers filmed a new cirrate ("hairy" or "fringed") octopus with a span of eight feet. The animal was spied using a diving saucer, the *Cyana,* near a hydrothermal vent 8,500 feet down in the northeastern Pacific. Cirrate octopods are named for the hair-like bristles surrounding their sucker disks. They also boast a unique pair of lateral fins that somewhat resemble ears. The new find is larger than any cirrate octopus ever caught, but may not be the biggest ever seen. The *Alvin* photographed a pink specimen in the Carribean in 1974 that may have been half again as large. In 1994, *Alvin* also spotted a new deepwater species. The world's press carried the tale of this seemingly confused mollusk, a male about fifteen inches long. It was attempting to mate with a much larger octopus—also a male.

The *Cyana's* crew spotted another new cirrate octopod in 1989. This cephalopod, also a Pacific denizen, displayed truly bizarre behavior. When poked by the *Cyana's* robot arm, the annoyed octopus gathered its arms together, trapping water in the membranes between them and taking on the appearance of a submarine pumpkin. This action is presumably designed to startle predators. It certainly had that effect on the French aquanauts.

A Parade of Mammals

As noted in the introduction, several new mammals are discovered each year. Such finds generally go unnoticed by the media unless the animal involved is large, strange, or adorable. However, each find is important to scientists, and some of the newest mammals are well worth meeting.

Take bats, for instance. Almost nobody likes these little animals, but they are marvels of adaptation, with their excellent flying ability and sophisticated echo-sounding apparatus.

Three new Asian bats were found in the period from 1939 to 1948 alone. Two of these are still known from the original specimen only, and have not been found again.

Australia is home to the white-striped sheath-tailed bat, an orange-brown animal with white stripes on its flanks. Despite its coloration, which is rather flashy by bat standards, the animal was only discovered in 1979 in northern Queensland.

The Australian region is best known, of course, for marsupials. Newer marsupial finds include the long-footed potoroo, a miniature kangaroo stuck with one of those endearingly silly-sounding names produced when Aboriginal terms are mangled by English-speaking "discoverers." This four-pound animal wasn't described until 1980, and is still known from just a handful of specimens collected in Victoria. The black dorcopsis wallaby, a member of another kangaroo-related family, was described from specimens collected on Goodenough Island in 1953 and has remained in hiding ever since.

The Proserpine rock wallaby is a gray-furred member of another mini-kangaroo group, the extensive wallaby family. The animal, which weighs just under twenty pounds, was collected for the first time in Queensland in 1976. Two years later, an even smaller species, the monjon, was discovered. Another wallaby, the warabi of Australia's northwest coast, had been added to the scientific rolls in 1963.

Dunnarts are gerbil-like marsupials with bodies three to six inches long, covered with soft, dense fur. Some species have adapted to desert climates by using their tails to store fat, the way camels use their humps. The Julia Creek dunnart was discovered in 1931, and only a few specimens are known from northern Queensland.

Closely related to the dunnarts are the ningauis, slightly smaller, longer-haired animals first identified in 1975. A captured ningaui adapted well, uttering a high-pitched cry when it wanted food.

The cinnamon antechinus is the largest member of its family of marsupial mice. This animal was described only in 1979. It lives in the forests of Australia's Cape York Peninsula. Like some other species of antechinus, it displays truly bizarre habits when mating season rolls around. The males burst into a fury of dominance fights and equally energetic matings, putting aside such mundane functions as eating. This frenzied activity continues until the exhausted, emaciated animals simply collapse and die. The male offspring born as a result of this outburst will live only until the next year, when they too will go out in a blaze of hormonal glory.

The mountain pygmy possum, a grayish little mouselike animal with a long prehensile tail, was one of the biggest surprises of recent years. It was known as a fossil from New South Wales, but had been extinct for 15,000 years. At least, everyone thought it extinct until 1975, when one was found checking out a trash bin behind a ski hut on Mount Hotham in Victoria. Since then this rat-sized "fossil" has been found in two other areas of Australia.

The New Guinea glider was found in 1980. This is the sixth known species of gliding opossum, the marsupial world's answer to the flying squirrels.

Eight years later, the Toricelli Mountains of Papua New Guinea produced the first specimen of the black tree kangaroo, also known as Scott's tree kangaroo. Australian biologist Tim Flannery had been searching for this animal since collecting an unidentified claw in 1985, and he continued his quest until he brought in a live specimen in 1992. This rare and highly endangered creature is the largest of the tree kangaroos, animals which have returned to the arboreal habitat the ancestors of today's kangaroos occupied thousands of years ago. The new species may be six feet long including the tail, making it Papua New Guinea's largest known mammal.

In 1987, the bronze quoll was described from the same island. Quolls, also called native cats, are weasel-like predators previously known from Australia and New Guinea. The newest quoll, described as having "a body color a deep bronze to tan brown with a black tail," is a resident of the low savannah country and is about two feet long overall.

In 1994, another new tree kangaroo was confirmed from New Guinea. Tim Flannery, the discoverer of Scott's tree kangaroo, reports that the newest find "is very primitive in its body plan and behavior." His point is that the animal is not particularly specialized for tree life. It does, in fact, spend most of its time on the ground. In size, it ranges up to four feet long with the tail and weighs over forty pounds when fully grown. The kangaroo is black with a white chest and other white markings. It shows no fear of people, emits a whistling call, and is quite endearing by human standards. It lives in mountain forests in the Indonesian half of the island. A formal description and scientific name are pending.

The opossums are the only marsupials to have maintained a presence in the New World. The black four-eyed opossum was described from Peru in 1972. The name derives from white patches over the actual eyes, which are black.

Hutias are ratlike animals of the Carribean, and three new species were described by the same scientist in 1970. All are rare and confined to small islands off the coasts of Cuba. A fourth new species, Cabrera's hutia, was first collected on another island in the mid-1970s.

Turning to the true rats, one might say that the last thing we need to do is discover new ones. After all, we spend a fortune trying to eradicate the rats we've already got. Be that as it may, they keep coming.

The Ilin Island cloud rat was found in 1981 on Ilin Island in the Philippines. It's one of the largest rats, with a total length of about eighteen inches. As in many members of the cloud rat genus, the animal's tail is not naked, but furry. Unfortunately, that 1981 specimen may have been the last of its kind. Ilin Island's forests have been destroyed, and a search in 1988 produced only one elderly man who recalled ever seeing the animal.

The Venezuelan fish-eating rat, equipped with broad feet and other adaptations for life around and in water, was found in the 1960s. The Mindoro rat is a brown rat whose long tail ends in a prominent tuft of fur. Three specimens from Mindoro in the Philippines were collected in the 1970s.

New species from the United States are very rare. Still, the diminutive Key rice rat (or silver rice rat) scurried around the mangrove marshes of the Florida Keys until 1973 without being noticed. Two were found in that year, and the species was formally described in 1978. A third rat turned up in 1979 and five more in 1980-81.

The small carnivores have their new members, too. The water

weasel of Columbia was described in 1978 from two examples found high in the mountains. It's about a foot long and sports dark brown fur on its back and a lighter shade on its belly. The Liberian mongoose was discovered in 1958. This animal is also mainly attired in dark brown, and a dark stripe runs along the side of the pale-furred neck. It is so little known that no photograph of it existed until Mark Taylor, a Canadian scientist, obtained one in 1990. Meanwhile, another mongoose had been described from Nigeria in 1984, and yet another from Madagascar in 1986.

It's amazing how long some interesting mammals have scampered about without attracting attention. New discoveries are sometimes right under our feet.

Voyage of the *Galathea*

The 1950-52 voyage of the Danish research ship *Galathea* resulted in much more than just the previously mentioned *Neopilina*. That a single expedition in the postwar era produced so many finds is a significant reminder of little we know even today about the abyss off our shores.

In the China Sea, the *Galathea* team, led by Dr. Anton Bruun, dredged up a new multi-tentacled sea cucumber. This is a round animal, colored a pale, translucent blue, approximately ten inches in diameter. The strange-looking invertebrate inhabits the abyssal waters below 11,000 feet. The same trawl brought up an oval sea cucumber, measuring about nine inches by six inches, which swims by means of a moving fringe running down its sides. This violet species was strange enough to require its own genus—suitably, *Galatheathuria*.

A trawl set 1,900 feet deep in the Tasman Sea produced four small fish, members of a new species of the bullhead family. This slender fish, with eyes and fins seemingly stuck on to a too-small body, went into the books as *Antipodocuttus galatheae*. The bullhead family, until this time, was thought to exist solely in the northern hemisphere.

Off Natal, the *Galathea's* efforts produced a new eel, *Coloconger,* so stout it didn't look like an eel at all.

The deepest and strangest new fish trawled up was *Galatheathauma axeli,* a Pacific find from almost 12,000 feet down. This species of angler fish, eighteen and a half inches long (very large for such a deep-dweller), is all black with a broad head. Resting on the bottom, it opens its cavernous jaws and lures prey in toward a large light organ dangling from the roof of the mouth. Curved teeth, looking almost like claws, line the edges of the upper jaw waiting to snap shut on a visitor.

The *Galathea* crew even found time to make a fresh-water discovery. On remote Rennell Island in the Solomons, they caught a sea snake whose ancestors somehow ended up in a lake and adapted to the change. The lakedwelling subspecies has darkened in color but otherwise remains very similar to its seagoing relations.

Dr. Bruun's only regret after this expedition was that "still no one has caught the Great Sea Serpent." He wasn't joking, but we'll come to that story in Section III. The *Galathea's* discoveries made plenty of zoological history, with or without sea serpents.

Kitti's Hog-Nosed Bat

How small can a mammal get?

As warm-blooded animals, mammals boast a complex thermal regulation system which requires a certain minimum size to function. Accordingly, there can never be a mammal the size of a housefly. (That includes Raquel Welch and her comrades in *Fantastic Voyage.*) Cold-blooded vertebrates can be smaller. The apparent record for all vertebrates belongs to the dwarf pygmy goby, a fish whose redundant name is longer than its half-inch body.

The current holder of the "world's smallest mammal" record

took over the title in 1973. In that year, Kitti's hog-nosed bat, also known as the bumblebee bat, was discovered by Thai scientist Kitti Thonglongya near his country's famous River Kwai. Thonglongya, however, died before he could publish a description of his find. A British colleague, John Hill, finished his work the following year, and the species' debut date accordingly appears in official records as 1974.

Not much over an inch long, the bat weighs a mere one-sixteenth of an ounce and has a wingspan of only six inches. Its face is strikingly ugly by human standards, but that didn't stop mammologists from swarming to study the circulatory system which allowed the bat to maintain a steady temperature in a body smaller than that of many insects.

The mini-bat, while only one of five new mammals described from Thailand since 1970, proved so unique that a new family was created to classify it. This makes the bat the lone inhabitant (so far, anyway) of the first new mammal family described in the twentieth century.

Today, the bumblebee bat is listed as one of the world's most endangered species. It lives in a handful of limestone caves now threatened by hydro-electric projects. No matter how small you get, it's hard to hide from progress.

Fishing in Siberia

New fishes are constantly being pulled from the seas of the world. We've already met the newest sharks. There are plenty of other examples, such as the eel-like *Lamprogrammus shcherbachevi*. This wide-ranging creature, over six feet long, was identified in 1993 from four specimens caught between 1972 and 1991 in the Atlantic, Indian, and Pacific oceans. Even more startling, due to its size and shallow-water habitat, was the discovery of a new stingray from

Thailand in 1987. The largest of the three specimens obtained measured over fifteen feet!

Discoveries in fresh water are rarer. A study of finds from 1944 to 1970 showed the number of freshwater species discovered annually dropped from 155 to 27, while discoveries of marine fishes declined insignificantly from 107 to 96.

That certainly doesn't mean we've classified all the freshwater types. In 1993, a diminutive new sculpin was identified in, of all places, the Potomac River in the eastern United States. However, finding a freshwater species today generally involves travel and even hardship. Just ask Russian fisheries biologist Mikhail Skopets.

Skopets is a specialist in the salmonoid fishes, which include trout, char, and others as well as the salmon. About thirteen years ago, his research took him to a remote area of Siberia, north of the Sea of Okhotsk. There locals described to him an even more remote spot, Lake El'gygytgyn, or White Lake, which held two kinds of char unlike those they caught in other waters.

Skopets was intrigued enough to make the trip. White Lake is actually an ancient meteoric crater six hundred feet deep. It is frozen ten months out of the year and is never far above freezing. After spending weeks huddled in a tent beside the lake, Skopets did indeed capture a new species, the smallmouth char, which he described in 1981.

The story didn't end there. Skopets also caught a twelve-pound specimen of a known predatory char in the lake, and in its stomach were bones that couldn't be identified. It seemed the local fishermen knew what they were talking about.

Skopets and four colleagues again crossed the icy tundra to the lake, 250 miles from the nearest village. Pulling in gill nets from a tiny inflatable boat in the midst of an Arctic windstorm, Skopets got his fish: a deepdwelling char so unusual it had to be placed in a new genus. The longfin char, as it was named, grew to a maximum size of thirteen inches. It is the most primitive char known, and brought new understanding of salmonoid evolution.

So new fish are still being found in salt and fresh water. But *on land*? Believe it or not, a red, inch-long Brazilian catfish discovered in 1984 has abandoned the water world. The fish is eyeless, scaleless, and, if put in water, immediately wriggles back onto the shore, where it breathes through its skin rather than its undersized gills. This bizarre sluglike creature is the only known fish adapted to living out of water nearly full time.

The Crested Iguana

The iguanas are an ancient group of lizards. They have evolved into a variety of forms, with one type, the marine iguana, even foraging under the surface of the Pacific Ocean.

In January of 1979, Fijian biologist John Gibbons visited the island of Yandua looking for a rare species, the banded iguana. He found something else: an emerald green lizard about thirty inches long, marked with narrow bands of white and further distinguished by conical spikes rising from its back.

The new tree-dwelling species was named the crested iguana. A notable characteristic of the handsome lizard is the transformation it undergoes if startled. The body color quickly darkens to black and the spikes stand erect, resulting in a much meaner-looking lizard, resembling a sort of reptilian Hell's Angel.

This obviously rare species existed only on Yandua, or so Gibbons thought until he discussed the lizard in 1980 with herpetologist Kraig Adler. Adler told him that an acquaintance had reported spotting an odd iguana in the movie *The Blue Lagoon*. Gibbons waited

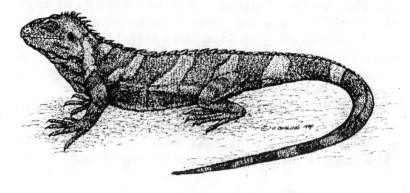

anxiously for the film to arrive in Fiji. When it did, he watched it closely, ignoring the charms of Brooke Shields in favor of reptilian extras. And there it was: a crested iguana, filmed in another island chain, the Yasawas.

Another recent discovery, a new monitor lizard from Yemen, also came about by accident. German zoologist Wolfgang Bohme was watching a television documentary on that Middle Eastern country in 1985. The program included a clip of a large native lizard, and the startled viewer realized that the filmmaker had unwittingly recorded an unknown species. Dr. Bohme and two colleagues flew to Yemen to track down the four-foot monitor. They didn't find it, but a second attempt led by Dr. Beat Schatti of Switzerland was crowned with success. The species was formally described in 1989, and six examples were given a place of honor in the Zurich Zoo.

Reptiles like this don't get the press that the more glamorous and cuddly mammals do, but researchers continue digging up new ones nonetheless. A biological survey of Cuba beginning in 1991 produced fourteen new reptile species, with five frogs and four fish thrown in as bonuses.

Other lizard finds included the 1964 discovery on the Carribean island of Virgin Gorda of a gecko that may well be the world's smallest reptile. The largest specimens were less than an inch and a half long including the tail. A sole specimen of another new gecko about the same size was collected from Haiti in 1966.

It's been some time now since a new crocodile was discovered. The family last welcomed a new species in 1935, when herpetologist Karl Schmidt described the Mindoro, or Philippine, crocodile. Some authorities initially dismissed this as just a local race of the New Guinea croc, but since Schmidt had discovered *that* reptile as well (in 1928), his opinion carried enough weight to establish his find. This freshwater beast, up to nine feet long, lived in small populations scattered across eight islands. It remains little-known and today is highly endangered.

A new crocodilian subspecies, the Apoporis River caiman, was described in 1954. Caimans are tropical alligator-like reptiles spread widely over Central and South America. The new type, limited in range to the Rio Apoporis of Columbia, was identified as a variant of the spectacled caiman. It has a very narrow snout and is colored a distinctive yellow-brown with blackish spots. Adults may approach seven feet when fully grown.

New turtles occasionally poke their snouts out of the water, as the narrow-bridged mud turtle did in 1965. Rare and still almost unknown, it hides out in the swamps and streams of Nicaragua, Panama, and Costa Rica. In 1984, a new species of the so-called side-necked turtles was found in southeastern Brazil.

The largest land tortoise in North America was only discovered in 1958. Scientists from the University of Illinois made the find when they happened on an unidentified shell being used as a chicken feeder in the Chihuahan Desert of Mexico. The bolson tortoise, as it's now known, may be eighteen inches long and weigh thirty pounds. Unlike other tortoises, which are solitary, bolsons live in communities with a definite social structure and as many as 100 inhabitants. They are endangered by hunting and development, and a captive breeding program is now underway.

The sea turtles were all classified in the last century, or so most herpetologists thought until 1967. In that year, three curious scientists took a fresh look at some strange specimens collected almost a hundred years before.

In 1880, an article published by Samuel Garman suggested that two unusual turtles from Malaya and Australia belonged to a new species. The leading herpetologists of the time rejected the idea, classifying the three-foot-long reptiles as aberrant green turtles. The question resurfaced periodically over the next several decades, but the green turtle identification remained predominant.

It didn't help that one of Garman's original specimens was shown without question to be a misidentified green turtle. The other was more peculiar, boasting a much flatter shell and differences in flipper design, scale pattern, and coloration, and that was enough to keep the problem from being entirely forgotten.

In 1967, herpetologists E. E. Williams, A. G. Grandison, and Archie Carr restudied Garman's specimen and a juvenile described in 1908. Their resulting paper established the flatback turtle as a distinct species. As their opinion was reinforced by observations of the living turtle from the beaches of northern Australia, more herpetologists agreed that an important discovery had been overlooked, and the flatback has been gradually creeping into the textbooks ever since.

The flatback has another distinction, a happy one: it is the only sea-turtle species the International Union for the Conservation of Nature (IUCN) does not classify as threatened in any way. It seems that one other difference between the flatback turtle and the green is that the flatback just doesn't taste as good.

The Rufous-Winged Sunbird

South American parrots weren't the only birds to emerge in the 1980s. The decade produced a total of forty-three new species of our feathered friends. This rate of about four species a year represents a surprisingly small decline from the six species a year being described before World War II.

One discovery came when ornithologist Flemming P. Jensen was in Tanzania's Mwanihana Forest, collecting bird specimens by the common device of stretching a fine mesh called a mist net between trees. Among the trapped birds was a colorful fellow about five inches long that Jensen described in 1981 as the rufous-winged sunbird. The sunbirds are a widespread group who fill the same flower-pollinating role in the Old World that hummingbirds do in the New. The new bird's plumage was metallic blue on the back and gray on the underside, adorned with a bronze throat and a red band on the breast. The bill is long and curves downward.

A return visit in 1985 produced more specimens and better delineated the bird's habitat. It turned out the sunbird was a creature of montane forests above 4,600 feet. The first specimen had been a stray, creating a lucky break for ornithology.

Campbell's Fairy Wren was discovered in the same way in 1980. Mist nets set on Mount Bovasi in Papua New Guinea produced two specimens of this previously unknown bird. Only a few others have been spotted since, and the species remains very elusive.

Bird discoveries have come from all over the world. Ash's lark was found in 1981 in southern Somalia. The next year, Cox's sandpiper was described from Australia. There remains a bit of uncertainty surrounding this bird: there are adult specimens, but no reports of juveniles or breeding grounds, so some ornithologists suspect it's a hybrid.

Tiny Amsterdam Island, in the Indian Ocean, produced a large and spectacular bird find in 1983. The Amsterdam albatross, as French ornithologists Pierre Jouventin and Jean-Paul Roux named their discovery, is a beautiful white bird with black markings. It was immediately listed as endangered: estimates of the population at its single isolated nesting site run as low as fifty individuals.

The Red Sea cliff swallow is known from a dead bird found on an island off Port Sudan in 1984. That carcass remains the only specimen so far.

Even the United States boasts a modern discovery. Hawaii's black-faced honeycreeper, also called the po'o-uli, was unknown until 1973. The stout little insect-eater lives only on Maui, where its population is estimated at about 140 birds.

Somalia recently produced another find, one that became the catalyst for an interesting and controversial story. The type-specimen of a new species of shrike with unique black and white plumage was collected by a British scientist named Edmund Smith. Instead of shooting his specimen, though, Smith trapped it alive. German ornithologist Jan-Uwe Heckel studied the bird, wrote the formal description, and then took the only known specimen of the Bulo Burti boubou shrike back to Somalia and released it into the wild.

On the one hand, Smith and Heckel had preserved a bird which might be so rare that its species could ill afford the loss of even a single member. On the other, some ornithologists objected that science can't be so tenderhearted: who knew what vital information might now go unlearned without a specimen to study?

We may or may not see that shrike again. We will, as a certainty, find more rare new birds as the effort to catalog the earth's species—and save them—continues apace.

Section II

Presumed Extinct

Introduction: Animals Lost and Found

When is an animal extinct?

This question is more difficult than it sounds. As we'll see, a surprising number of creatures have been declared extinct by some authorities (sometimes by virtually everyone) and yet have resurfaced. Many more remain in twilight, their fate uncertain.

The International Union for the Conservation of Nature (IUCN) defines an animal as *extinct* when "there is no reasonable doubt that the last individual has died." The next level is *extinct in the wild*, meaning that a domestic population exists but "exhaustive surveys...throughout its historic range have failed to record an individual." Then comes *critical,* defined as "facing an extremely high probability of extinction in the wild in the immediate future." One criteria for this classification is "Population estimated to number less than fifty mature individuals."

Difficulties in classifying an animal may arise due to its remote habitat. Steller's sea cow (normally considered extinct) has probably been hunted out of existence, but questions will remain until someone undertakes a thorough survey of the desolate Arctic waters where the animal could have survived.

In other cases, an animal's habitat was not adequately charted: many an "extinct" creature later turned out to be living in another location or even on another land mass. New Zealand's takahe, a large flightless bird, played this game well, popping up three times on different islands or in different areas. Finally, some animals are just plain good at hiding. The Eastern cougar of the United States may be truly extinct, but the cat is so stealthy that any remnant population might be found only by accident.

Some of the animals in this section, like the dusky seaside sparrow, are probably gone for good. With others, such as the cougar, the chance for rediscovery is slim but real. In the most hopeful cases, an "extinct" animal's survival is beginning to look probable: the thylacine of Tasmania is apparently still clinging to a precarious existence. There are examples of cases where such hopes have come true. A snake-hunting raptor, the Madagascar serpent eagle, was found in 1988 after a fifty-eight-year "extinction," and the Tonkin snub-nosed monkey was rediscovered in Vietnam in 1992.

Extinction isn't always the fault of humanity. "Normal" extinctions, driven by competition or climatic change, are an eternal part of life on this planet. But there is a growing consensus that for humans to drive an otherwise-healthy species into extinction, either deliberately or through neglect, is ecologically and morally unacceptable. The rediscovery of a supposedly vanished species is an occasion for rejoicing: a sign that, no matter how badly we have managed the earth's ecosystem, we sometimes get a chance to restore a little piece of it.

Steller's Sea Cow

The known history of Steller's sea cow is a brief and tragic tale. In 1741, naturalist Georg Wilhelm Steller was shipwrecked on Bering Island. This is one of the Komandorski Islands, which lie in the frigid waters between Kamchatka and the Aleutians. There he and his companions met the sea cow. It was a huge plant-eating mammal, up to thirty-five feet long, with a bilobed tail like a whale's and a placid disposition that made it easy to approach—or harpoon. The rough-skinned creatures were also loyal, gathering around an injured animal. That instinct, while admirable, did little to promote their survival.

After Steller's crew finally returned to civilization, sealers and other voyagers began stopping off in the sea cow's haunts to slaughter the inoffensive mammals for their meat. By 1768, the species had apparently been hunted to extinction.

Evidence for the animal's continued survival is slender but per-

©C.GOSLING 94

sistent. Native hunters reported killing them as late as 1780. Early Russian colonizers of Bering Island reported sighting sea cows in the 1830s. Fifty years later, the explorer Nordenskiold returned from the region with a sea cow skeleton of unknown age and a tale of a live sighting from 1854. In 1910, fishermen in Russia's Gulf of Anadyr reported a sea cow stranded on the beach, but the report was never investigated. The sea cow's range may well have extended beyond the Komandorskis, as eighteenth-century reports placed them along the Kamchatkan coast and occasionally in the Aleutians.

Despite these unverified if intriguing tales, the commonly held scientific opinion is that the unfortunate sea cow didn't make it out of the eighteenth century. All that remained were a few sketches, some incomplete skeletal remains, and the sea cow's smaller relatives from warmer climates, the dugong and the manatee. Both of these latter animals, it should be noted, are in danger of extinction themselves.

A Russian whaler, the *Buran,* revived interest in the sea cow story in 1962. In the same gulf where the 1910 report had originated, the *Buran*'s crew sighted a herd of dark-skinned animals over twenty feet long they could not identify. The description, provided by experienced seamen who observed the creatures at a range of less than a hundred yards, matched Steller's sea cow very well. Another sighting by Russian fishermen was reported in 1977.

Unfortunately, no one has seen these alleged sea cows or any others since. A Russian mammologist, V.G. Heptner, has dismissed the recent reports as misidentifications of female narwhals, which are almost the right size and do not possess the distinctive tusks of narwhal males. Did the Russian witnesses see only whales or some other known sea mammals? Or did they find the last survivors of a species presumed extinct?

It must be admitted there's no hard evidence the sea cow is still alive. But it's fair to ask: wouldn't whalers, of all people, know whales when they saw them? And it's only human to hope.

Tasmania's Elusive Tiger

The thylacine is, or was, a doglike marsupial of Australia and Tasmania. Because of the broad, dark brown stripes marking its lighter brown coat, it was also known as the Tasmanian tiger. In build it resembled a greyhound or a lean wolf, with the average adult weighing about fifty-five pounds. The Australian mainland variety apparently died out perhaps three thousand years ago, but the animal prospered on Tasmania until sheep farmers arrived.

The newcomers quickly placed bounties on the carnivore's head, and the thylacine was hunted relentlessly. The last known specimen, "Benjamin," died in a Tasmania's Hobart Zoo in 1936.

However, reports of live thylacines keep trickling in. By one

count there have been sixty-four sightings since 1936, although not all are reliable. Dedicated naturalists have spent years tracking down witnesses and casting tracks. A 1945 expedition led by David Fleay collected tracks, droppings, and a tuft of hair left by a thylacine which entered a box trap but quickly backed out. (Fleay held a rather ignominious distinction: he was taking photographs in Benjamin's cage in 1936 when the last known thylacine bit him on the behind.) The *Guinness Book of Records* continues to list the creature under "Rarest Mammals." The editors cite a close-range sighting by park ranger Hans Naarding, who spotlighted a thylacine in northwestern Tasmania in 1982.

Eric Guiler, former chairman of Tasmania's wildlife protection board, notes that the island still has large areas of unpopulated land suitable for thylacines. In his book on the animal's sad history, he lists evidence including clear thylacine tracks found in 1958 and 1960 and several animals killed in the thylacine's distinctive manner (only the internal organs are eaten) in addition to the live sightings. A thylacine was apparently killed in 1961, but in a bizarre turn of events, the carcass was stolen to sell for profit, then presumably dumped at sea when no sale of the illegally possessed animal could be arranged. In another incident, four men reported they had trapped a live thylacine, but they released it and kept quiet for years afterward because they had been poaching at the time.

American researcher Jim Sayles led two efforts in the late 1970s, with results totaling one clear set of tracks, one thylacine call, and one achingly brief glimpse of an adult animal by moonlight. (The thylacine has a unqiue cry, described as resembling the creak of a door.) Dr. Guiler concludes the species is still around, although extremely rare and certainly endangered.

In 1983, U.S. broadcasting maverick Ted Turner offered a $100,000 reward for proof of the thylacine's survival. Guiler feared this would lead mainly to a lot of unqualified people tromping around in Tasmania's rugged wilderness, and the government reminded prospective thylacine hunters that anyone who trapped or shot a thylacine might get the reward but could also get six months in jail. To date, neither the reward nor the thylacine have been collected.

There are reports of thylacines surviving in remote areas of the Australian mainland. A photograph of an alleged thylacine from New South Wales was published in 1976. However, this picture showed the animal's tail curled over its back. Thylacines have a very inflexible tail that can't be curled or wagged like a dog's.

More attention was paid to pictures taken in 1984 by a tracker named Kevin Cameron. These photographs, allegedly made in Western Australia, show what looks like the distinctively marked hindquarters and tail of a thylacine. The tail stands out stiffly, as it should. Many authorities doubt the photos' authenticity, and everybody wishes Cameron had a picture of the head. Still, the episode renewed interest in the animal's fate.

The focus shifted back to Tasmania in 1995. On January 25th, park ranger Charlie Beasley observed a thylacine "about half the size of a fully matured German shepherd dog" in a forest on the northeast region of the island. The sandy-colored animal was in view for about two minutes.

A live or recently dead thylacine has yet to turn up, but it would not be a great shock to the zoological community if one did. That one *might* is a very pleasant thought.

The Takahe

No place on Earth boasts a variety of bird life to rival New Zealand's. With no mammals to rival them, New Zealand's birds sprouted into a dizzying variety of forms great and small. Even after the arrival of Europeans, this avian kingdom was slow to reveal its secrets. Some mysteries are unsolved to this day.

New Zealand boasts three known species of the comical featherball called the kiwi bird. Maori lore includes a fourth, much larger, species of kiwi, which they called the roa-roa. It has never been found by science, but there does exist a cloak of kiwi-like feathers which are much larger than those of any known kiwi.

The takahe was another flightless bird. It was turkey-sized and

boasted striking red, green and blue feathers. It, too, was considered extinct by the time Europeans settled New Zealand. Only fossils were known.

Accordingly, the shooting of a takahe in 1849 was a startling event. It took place on Resolution Island, which is just off the coast of South Island. The bird was not seen again until two years later, when a second specimen was taken on another island.

The takahe then disappeared again, this time for 28 years. Then another one was killed, this time on the mainland of South Island. Incredibly, the bird dropped from sight a third time, until a lone takahe was caught in 1898. Once more the takahe was crossed off the extinct list, then scribbled in again when no one could find another specimen.

It wasn't until 1948 that the takahe defiantly reappeared. In an isolated valley of South Island's Murchison Mountains, Dr. Geoffrey Orbell, a physician with a fortunate curiosity concerning the missing bird, led an expedition that netted two takahes. Orbell discovered that previous searchers had overlooked, not a few stragglers, but a viable breeding population of at least ninety birds!

Today, the takahe continues its comeback. Wildlife Service workers are trying to eradicate egg-eating stoats, which were introduced deliberately into New Zealand to control the rats and mice previously introduced by accident. With this assistance, plus a protected reserve and a captive breeding program, the takahe population has steadily increased. If ever a species showed how resilient nature can be, it is this unique bird, which can't fly an inch but certainly knows how to hide.

Most spectacular of New Zealand's inhabitants were the moas, among the largest birds ever known to have lived. It appears these flightless wonders, some species of which stood eleven feet tall, were already in decline when humans first arrived on the islands. The early Maoris may well have roasted the last giant moa over a campfire.

Then again, perhaps not. No one can be certain just when the moas died out. It is interesting to speculate whether the presumably extinct roa-roa may have been a small moa. Maori reports speak of such a bird being hunted as late as the mid-1800s.

The most tantalizing, albeit unproven, possibility is that some moas may still roam the New Zealand forests. Sightings were reported in 1928, 1940, and 1963, although none resulted in photographic or physical evidence.

In 1993, three people hiking in the Cragieburn Range on New Zealand's South Island reported spotting a live moa, forty meters away, clad mainly in reddish brown plumage and standing about six feet tall. One witness, a former military survival instuctor, had a camera. He claimed he chased the startled bird and took one picture on the run. The photo shows what could be the outline of a large flightless bird, but the image is too fuzzy, even after computer enhancement, to be identified.

Dr. Ken Hughey of the New Zealand Department of Conservation commented, "We can't know yet if it's an emu or a moa, but we are very interested." The presence of an emu, a native of Australia, would be a mystery in itself: there is an emu breeding colony in captivity on South Island, but its manager reported no escapes. The Department initially planned a search of the sighting area, but called it off "pending further investigation." Private searches have been unsuccessful.

There the matter rests, awaiting physical evidence, or, at least, a better photo. Could there still be moas? One report is not proof, but speculation is irresistible.

The Eskimo Curlew

North America has offered avian puzzles of its own. If ornithologists put out a "Most Wanted" list of missing birds, the Eskimo curlew would have been a fixture for almost twenty years.

The curlew, a migratory bird just over a foot long, was once abundant. Unfortunately, it was also tasty. The birds were slaughtered for food in enormous numbers. Most were shot or netted in the central United States as they migrated from their winter homes in Argentina to nesting sites in the Arctic. Curlew flocks had a fatal and seemingly inexplicable habit of rising when they were shot at only to settle down again in the same or a nearby spot.

By the early 1900s, the massacre of millions of birds had reduced the curlews to a fraction of their former numbers. But the killing continued, until by 1930 it was hard to find a curlew. Reports of curlews migrating in pairs or alone were made in Texas in 1945, then again in 1959 through 1962.

For years after, no one reported an Eskimo curlew. The bird was almost universally consigned to extinction. As an epitaph, there was even a book and an animated TV special, *Last of the Curlews,* chronicling the loneliness of the very last bird.

Accordingly, the curlew's reappearance constitutes one of the most surprising comebacks in recent years. A pair were spotted in Texas in 1981. Canadian Wildlife Service personnel found an active nest in 1987, the first such nest reported since the American Civil War era.

A 1990 estimate put the population at 100 to 150. The plucky bird is still endangered, but endangered is much better than extinct.

The Eastern Cougar

The cougar (or panther, puma, painter, catamount, mountain lion, or whatever) is one of North America's most magnificent animals. It is also a species which has been trapped, shot, and poisoned almost out of existence.

Today, the western subspecies seems to be holding its own and has begun a modest comeback. The Florida panther has been reduced to a dwindling, inbred population so small (probably under fifty animals) that it may not be genetically viable. In other words, even if humans don't kill a single animal from now on, the cat may not survive in the wild. Captive breeding is probably the only hope.

The once-numerous Eastern cougar (*Felis concolor cougar*), which had the bad luck to live where the most humans chose to settle, was even less fortunate. Its extermination was for over a century a matter of government policy, and states placed bounties on the animal. Expert opinions on when the subspecies became extinct ranged from 1910 to 1946, with a specimen trapped in Maine in 1938 sometimes considered the last of its kind.

Was it? Perhaps not. In 1958, a biologist observed a cougar in his headlights in New Jersey. Over the next twenty-five years, more than 300 sightings in the U.S. and Canada were reported to various authorities. One of these cases involved a bold female who left her tracks around the outskirts of Montreal in 1959. A panther of uncertain origin was killed in Tennessee in 1971. Tracks, hair, and droppings found in New Brunswick in 1992 were identified by that province's wildlife authority as belonging to a panther.

Eventually, the evidence grew strong enough to put the cat on the U.S. endangered species list. Tracks and droppings found in 1981 were enough to persuade biologist Robert Downing, the author of the Fish and Wildlife Service's cougar recovery plan, that at least one cougar roamed the national forests of the Virginia-West Virginia area. Based on eyewitness reports, Downing pinpointed the mountains of North Carolina as another promising location. Finally, in 1994, droppings identified by state wildlife officials as cougar were collected after three animals were seen in Vermont.

Just the possibility this lithe predator still roams the Eastern forests is enough to lead some researchers to devote years to tracking down reports. One private effort, by the Eastern Puma Research Network, logged 245 sightings in 1994 alone.

The Eastern cougar's current existence is not a certainty. Many sightings prove to be mistakes involving dogs, bobcats, or other animals. It's also possible that an occasional western cougar is released by or escapes from people who keep them as dangerous and usually illegal pets.

A final problem is that many reports (thirty-nine out of the 245 mentioned above, for example) involve claims of black panthers. While black panthers have been killed on very rare occasions in Central and South America, there is not a single proven case of such a variation from our own continent. (There is some confusion of terms here: black leopards, found in many zoos, are sometimes referred to as "black panthers.")

All this will probably keep the majority of scientific opinion on the "extinct" side until there is a good film or a live or recently dead specimen to examine. Cougars are legendary for their ability to stay hidden, though, and the question may hang unsettled for years to come.

And I think in this empty world there was room for me and a mountain lion,

D. H. Lawrence once lamented. Given all the species we have eradicated from this continent, it's at least a little reassuring to think there may still be room for this shadowy cat after all.

The Survivor Seals

Those who believe that some species presumed extinct are still with us can take comfort from the case of the Guadalupe fur seal. This large mammal (males can weigh over 300 pounds, while females are only a third as heavy) was a common resident of the Mexican and California coasts before European settlers came. Intensive hunting apparently wiped out the dark gray seal by the 1850s. The species was found alive a few years later, and of course was immediately hunted again until its second presumed demise in the 1870s.

No one saw another Guadalupe fur seal until 1928, when two bulls were captured by a fisherman who sold them to a zoo. When there was a dispute over his payment, the man returned to the herd he'd found and killed the rest of the seals. This barbaric act apparently consigned the species to extinction once more.

Sightings in 1949 and 1951 gave conservationists some hope, but it wasn't until 1954 that a small herd—fourteen animals—was found on Guadalupe Island. This time, at long last, they were protected.

The closely related Juan Fernandez fur seal managed a similar reappearance. Declared extinct in 1917, it was found again on the

Juan Fernandez Islands off Chile in 1968. Today there are several hundred seals living in this protected sanctuary.

The Caribbean monk seal apparently isn't so lucky. In 1494, Columbus took note of this animal's abundance. By 1911, humans had found and slaughtered the last known herd near the Yucatan peninsula. The last definite record was from 1952, although sightings of one or two individuals were reported in 1964 and 1969. A report of two seals off the southeastern Bahamas in 1974 might have marked the last time this species was encountered. A 1980 expedition to this area produced no sightings, and it is quite possible the species is now extinct.

The same fate may have befallen the Japanese sea lion. This animal is usually classified as a race of the California sea lion, although recent study of remains indicates it is—or was—most probably a full species. By 1951, this once-plentiful sea lion was reduced to a known population of fifty or sixty animals on the island of Takeshima. That group has since been exterminated, and there may be no others.

Sometimes, even the most efficient killing machine in the animal world—man—overlooks a remnant of his prey. The Guadalupe and Juan Fernandez fur seals' success in hiding for decades in the vastness of the sea gives us some hope that the monk seal and other species may have done the same.

The Dusky Seaside Sparrow

Several once-numerous American birds have come to an ignominious end, with the last known specimen dying of old age in a lonely cage in a zoo. Such was the fate of the passenger pigeon and the Carolina parakeet, both of which officially passed away in the Cincinnati Zoo in 1914 and 1918 respectively.

The dusky seaside sparrow may have died out in the same manner. A victim of development and mosquito control efforts that eliminated its marshy habitat on Florida's east coast, the dusky's struggle for survival was played out in the shadow of the ultimate symbol of progress, the Apollo moon launches.

The bird's numbers had been dwindling for a long time before a wildlife refuge was established in 1971. Even then, fires and mosquito management efforts continued to shrink the suitable habitat, and bird counters logged fewer sightings every year.

Ornithologist Herb Kale, who led the effort to save the dusky, felt the bird's fate was sealed when, in 1973, it was reclassified from a species to a subspecies. Much more effort is likely to be spent on an animal if it's considered a species, and therefore unique. In addition, bird watchers, an important constituency, lose interest, because subspecies don't count on a birder's "life list." Genetic analysis has subsequently confirmed the dusky was a subspecies, but that doesn't diminish the sadness of those who tried to preserve it.

In 1979, in a last, desperate effort to save the dusky, ornithologists captured five of the six birds they could find. All, including the one who eluded capture, were males. Unable to find a dusky female, the rescue team tried to at least preserve the bird's genes by crossbreeding the males with their closest relative, the Scott's seaside sparrow. Ironically, this was done in a facility provided by a symbol of development, Walt Disney World, in a gesture mixing altruism and public relations. (In fact, Disney originally proposed that the dusky-Scott's hybrids be designated a new subspecies called *Ammospiza maritima disnei*.)

The last known dusky seaside sparrow, "Orange," died on June 16, 1987. Two years later, a storm damaged the roof of the research compound. The four living dusky-Scott's hybrids, which varied from 50 percent to 87.5 percent dusky, died or escaped. That, at least, was Disney's story. Author Mark Walters reports that rats actually got into the cage and killed at least two birds. One or two sparrows escaped or were released, and the storm was a cover story to keep Disney from appearing negligent.

Whatever the truth in this affair, it is possible that the escaped hybrids have mated, and that the dusky, in some form, is not entirely gone. It's also possible—although just barely—that some pure duskies have escaped searchers. There are bird species which have eluded detection for many decades. If, however, this little bird is truly extinct, then the moment of its passing is known with rare and saddening precision.

Found and Lost:
The Ivory-Billed Woodpecker

A more ambiguous and possibly even more tragic story involves the magnificent ivory-billed woodpecker. This bird is the largest U.S. woodpecker, over two feet long and handsomely decked out in formal black and white plumage. One admiring expert, ornithologist Dr. Lester Short, wrote that, "If the woodpecker world had royalty, the ivory-bill would be king."

The ivory-bill developed a very specialized diet, relying on beetles living in dying or recently dead trees in Southern forests. When lumbermen began clearing those forests, the bird could not adapt.

The ivory-bill was officially last recorded in the U.S. in 1950, when a pair was spotted in Florida. Ornithologists have searched the Southeast ever since, pursuing scattered reports of calls and sightings but never proving the bird's survival. The last piece of hard evidence, a single feather, was collected after a Florida sighting in 1969. The latest evidence came from Dr. Jerome Jackson, who played recordings of ivory-bill calls throughout the bird's range. In 1987 and 1988, he received replies, but failed to see a living bird. No one else did, either, and the Fish and Wildlife Service pondered whether to remove the ivory-bill from the list of endangered species on the grounds of extinction.

The ivory-bill also lived in Cuba, where it was sometimes considered a different subspecies, based on a minor difference in markings. By the 1970s, the Cuban ivory-bills were also widely considered extinct.

One of the great discoveries—or rediscoveries—in modern ornithology came on March 16, 1986, when a Cuban team spotted the first conclusively identified ivory-billed woodpecker in at least fifteen years. Short and other American ornithologists confirmed the bird's continued existence. An expedition invited by the Cuban government spotted two or possibly three individuals in 1987.

The Cuban population was clearly a small and highly endangered one. The patch of forest it inhabited was rapidly being logged out of existence. The logging was halted as a conservation measure, but that action now appears to have been inadequate or just too late to save the woodpecker. In 1993, a three-month search found not a single bird. A spectacular species had been lost, found, and was now lost once more, perhaps for all time.

Will the ivory-bill be rediscovered again, either in Cuba or the United States? Ornithologists can only search and hope.

One final note: the limbo inhabited by the ivory-bill is, unfortunately, shared by the ivory-bill's closest relative. The Imperial woodpecker of Mexico has not been seen for certain since 1958. The largest of all the world's woodpeckers, it, too, may be extinct.

The Sinai Leopard

Leopards are highly adaptable cats. They range across Africa and much of southern Asia, occupying all types of habitat. Leopards have even been seen lounging on carport roofs in Nairobi, seemingly unconcerned about the nearness of man. With its flexibility, stealth, and nocturnal habits, the leopard has survived the pressures of human development better than any other big cat.

One subspecies, the Sinai leopard, seemingly adapted to nature but not to man. These cats developed lighter spots for concealment in the desert, and switched from being night to day predators to match the habits of their chief prey, the native goats called ibex. After World War II, though, the leopard was apparently driven to extinction by hunting and the pressure of increasing human settlement.

Fortunately, not all the land was developed. Even under the pressures of population growth and war, the Israeli government pursued a policy of creating wildlife reserves. As the ibex became concentrated in the reserves, it was natural that any predators would follow them. But the ibex's only predator was the leopard, and everyone knew there were no leopards.

By 1971, Israeli zoologists were beginning to wonder. Ibex were being killed by predators which removed the entrails before eating the carcasses—a habit unique to leopards—and left leopardlike tracks as well. Game wardens began leaving food for the unseen marauders, hoping to lure them within camera range. The food was accepted, but it was taken with such cunning that four years passed before one of the sly freeloaders was finally photographed in the Judean Hills, and the Nature Reserve Authority announced the Sinai leopard's rediscovery.

The leopard's return is a testimony, not only to the animal's natural skills, but to its grit and determination to survive. Reduced in numbers, driven from the best parts of their range, the few remaining cats retreated to the most inhospitable deserts of the region and somehow hung on. As the ibex gathered together and increased in number, so did the leopards.

Today, the subspecies is still considered threatened, but its population is on the increase. The story is a tribute to successful wildlife conservation—and to one of the world's most resilient predators.

The tale of the Sinai leopard no doubt gives hope to those believing in the survival of America's previously discussed Eastern cougar, so does the tale of the Corsican wildcat.

This slender, dark-furred subspecies of the African wildcat was described from a lone specimen in 1929. Two skins were obtained shortly after this, and then the cat dropped out of sight. While not much larger than a house cat, the wildcat was confined to this modest-sized island and should hardly have escaped notice. But it did, disappearing completely for nearly sixty years. It was 1986 before the discovery of two new specimens resurrected the Corsican wildcat from its presumed extinction.

The mysterious nature of the cat, it seems, is more than a matter of legend.

The King Bee

In 1859, famed naturalist Alfred Russell Wallace discovered a new species of bee in the Molucca Islands. It was black with white markings and boasted enormous mandibles, or outer jaws. Its most striking characteristic, though, was its sheer size. At more than an inch and a half long, it was easily the largest bee in the world. Not surprisingly, it became known as the king bee (even though both specimens Wallace collected were females).

The world's largest, most conspicuous bee then vanished from science's field of view. Wallace's two examples remained the only known king bees.

The species wasn't rediscovered until 1981, when Dr. Adam Messer, then a University of Georgia graduate student, went in search of the bee and found it. Further study established that the bee lived on at least three islands in the chain, including the one where Wallace first encountered it. Somehow, this unmistakable insect had been overlooked for more than a century.

The king bee's size is not its only peculiarity. For one thing, it doesn't make honey. For another, the bee builds its nests in trees, inside the nests of termites who have conveniently bored out the wood. (The bees line their nests with a resin the termites can't chew through.) The king bee is a subletter—or perhaps, from the termite's point of view, merely a squatter.

Even more humbling for the "king," it is only the female of the species that's enormous. The male is about half her size. That's still pretty large by bee standards, but it makes one wonder whether the species has been misnamed as well as misplaced.

The Huia

In addition to the giant moas, New Zealand also produced some interesting "normal-sized" birds. For example, the kakapo is the world's largest parrot, weighing up to seven pounds. It is also the only flightless member of the family (it does glide a little). The kakapo was considered extinct or practically so by the 1970s, but turned up again on Stewart Island after several years of uncertainty. It's odd that the bird was missed, since the male lets out a booming mating call for hours on end that is loud enough to be heard three miles away.

Even more unusual was the huia, a black-feathered bird about the size of a crow. It had orange wattles on both sides of its face and a long, broad tail, the feathers of which were tipped with white.

The most unique feature of this species was its bill, or rather its bills. The huia was the only bird in which the male and female had significantly different types of bills. The male's was straight, the female's sharply curved.

The bills interested scientists the most, but the tail feathers were the huia's undoing. They became prized fashion items for both Maoris and Europeans, meaning that practically everyone in New Zealand was hunting the unfortunate bird. Eventually the hunters "won," despite an appeal from Maori chiefs in 1892 that persuaded the island's governor to ban hunting of the bird.

The huia was last definitely spotted in 1907, although searchers reported its call a few years later. The International Council for Bird Preservation (ICBP), the leading authority on such things, classifies the bird as having gone extinct circa 1912.

Ornithologist William J. Phillipps wrote a book about the bird in 1963, however, which listed twenty-three sightings made after the official extinction date. He accepted a 1961 sighting as genuine and concluded the bird still existed. Interestingly, the post-1907 sightings, taken as a whole, indicate a shift in the bird's habitat. The survivors appear to have been those on the northern fringe of the old range, and to have retreated from human intrusion and established themselves in less inhabited territory. That, at least, is the most hopeful possibility.

The Tarpan: Back from the Dead?

We have dealt with animals which returned from a presumed state of extinction. But is it possible for an animal to come back from *definite* extinction?

The case in point is the tarpan, the wild horse of Europe. A compact gray horse with a long coat and bristly black mane and tail, it roamed Europe and western Asia. One variety, the forest tarpan, lives on in a way in Poland, where forest preserves are maintained for herds of ponies of mostly tarpan blood. The main type, sometimes called the steppe tarpan, vanished through hunting and interbreeding. The last purebred tarpan died in the Ukraine in 1876.

In the 1930s, two enterprising German zookeepers, brothers Lutz and Heinz Heck, attempted to bring the tarpan back. The tarpan was then considered the ancestor of all European horses (today, the ancestry picture is considered more complicated). Accordingly, the Hecks reasoned, the tarpan's genes still existed in its descendants.

A program of selective breeding began at the Berlin and Munich Zoos. The Hecks mated what they considered the most tarpan-like horses, then matched those offspring that showed tarpan characteristics. After several generations, they had animals which certainly looked like tarpans, and whose offspring bred true to type.

Today, several zoos house examples of this "reconstructed" tarpan. But are the animals really tarpans? The breeders said they were: some zoologists consider them only replicas, a kind of walking Xerox of the real thing. There is debate about whether the Heck brothers even started with a good choice of tarpan relatives in the first place. (No one, apparently, has obtained DNA from preserved tarpan skins and compared it to the living animal.)

Whatever position one takes, the attempt to bring back the tarpan is an interesting experiment. Perhaps it's most accurate to list the tarpan as extinct...but with an asterisk.

The tarpan was not the first animal the Hecks had tried to revive. Around 1900, they attempted to rebreed the extinct European wild cattle known as the aurochs. Lutz Heck crossed Corsican cows and Spanish bulls to bring out the aurochs characteristics. Twenty-five years later, he got his first "aurochs," complete with the distinctive spiral horns of the breed. His brother, working at the Munich Zoo, started from different stock, including West Highland cattle and the gray cattle of the Russian steppes, and produced a similar strain. Some were reintroduced to the wild, where they became more aggressive and more distrustful of people. Today, there are some two hundred of these animals, although they are generally not accepted as "real" aurochs.

Other breeding-back attempts have involved the Atlas lion, the Caucasian bison, and the hemippus, or Syrian wild ass.

In 1982, a report circulated in the world's press that Russian scientists had recreated a species of mammoth, cloning it from DNA taken from frozen specimens. Unfortunately, this most fascinating of all animal resurrections was a hoax.

Could it ever be done? In theory, resurrecting, say, a tyrannosaurus from fossil DNA is possible. In practice, *Jurassic Park* is likely to remain fiction. Even if we could successfully regenerate living animals from DNA, which has never been done, the odds against recovering the entire undamaged genetic code for an extinct animal are enormous. A pioneer in such research, geneticist Russell Higuchi, compared the problem to reassembling, in the dark, a shredded encyclopedia written in a foreign language—without using your hands.

Przewalski's Horse

The eastern relative of the tarpan is worth mention in its own right. This is Przewalski's horse, which some believe belongs to the same species as the tarpan, and which is—this is a bit difficult to sort out—the only "pure" wild horse left in the world, *assuming* there are any wild ones left.

Like the tarpan, this former resident of eastern Asia is small by horse standards, standing less than fifteen hands (sixty inches) high and weighing under 700 pounds. It is normally brownish or yellow-gray, with a paler nose and underside and darker mane, tail, and shanks. A dark dorsal stripe runs from the stiffly erect mane back to the tail. Also like the tarpan, it is known as a spirited animal which sometimes drove away or killed much larger domestic stallions and then bred with the mares. The status of Przewalski's horse in the wild is uncertain. The last one captured was "Orlitsa," a mare taken in

© 1994 C. GOSLING

1947. A group of seven were sighted by scientists in 1966, and a few may still survive along the Chinese-Mongolian border. It's unknown whether the occasional reports of wild horses from this desolate region concern true Przewalski's horses or animals of mixed ancestry.

Over a thousand pure examples of the species do survive in zoos around the world. All are descendants of a single small herd sent to Europe before World War II.

United Nations Development Program researchers have produced a plan to reintroduce the horse to the Mongolian wilderness. The idea calls for assembling the most genetically diverse herd possible from the captive population. The next step will be to turn these horses loose in a large enclosure in Europe so the scientists can learn about the dynamics of a Przewalski herd in a free-ranging environment. When such data as the right mix between the sexes has been gathered, the herd will be reshaped accordingly and only then plunked down intact in its traditional range. The most intriguing question is whether these newcomers will find any old friends waiting to greet them.

Fraser's Dolphin

In 1895, a dolphin's body washed up on the beach in Borneo. The skeleton passed into the collections of the British Museum, where it rested, unstudied, for six decades.

Thus it came about that Fraser's dolphin is one of the newest cetaceans known to science, having been unearthed and described only in 1956. Between the collecting of that first skeleton and its official description, there were no reports at all concerning the animal.

That's surprising, because Fraser's dolphin is neither small nor hard to distinguish. It grows to be more than eight feet long and may weigh over four hundred pounds. The back is black, and three stripes run along the sides: gray, black, then gray again. The belly is white.

We know all this because Fraser's dolphin finally turned up again. On January 27, 1971, a fishing boat in the eastern Pacific hauled in no less than twenty-five specimens of the missing cetacean. Three weeks later, two more were caught off South Africa. Two weeks after that, another appeared on a beach in Australia. It was a widespread and spectacular reappearance for an animal no one had seen in decades.

Since then, schools containing hundreds of dolphins have been seen, and one estimate is that there are 136,000 just in the eastern Pacific. The question concerning Fraser's dolphin is not "Where is it?" but "Where was it for so long?"

Fraser's, however, is not the only dolphin to have disappeared for long decades into the vastness of the sea.

Heaviside's dolphin, also known as the Benguela dolphin, is a little cetacean less than five feet long from South African waters. It lacks the usual dolphin beak, and, except for having a short dorsal fin, is colored and shaped like a miniature killer whale. Heaviside's dolphin is not new, being described in 1828: however, no one saw it again until 1856, and then it vanished for over a century. The species wasn't observed again until 1965, and a proper description was published only in 1988. Nothing is known about its range, habits, or numbers.

A slender long-beaked skull found at an unrecorded location was described as the "clymene dolphin" in 1850, but most experts dismissed it as an abnormal example of the known spinner dolphin. That remained the case until 1977, when Dr. William Perrin matched the old skull with more recent remains and with two puzzling female dolphins caught alive in Florida in 1965. The result was the belated rediscovery and acceptance of the clymene (or helmet) dolphin, which does resemble the spinner but has a shorter beak, thicker body, and fewer teeth. It remains almost unknown, which is likely to be the case when you're mistaken for somebody else.

The pygmy killer whale, a round-headed, beakless dolphin which may be over nine feet long, was recognized from one skull found in 1827. The actual animal wasn't seen until 1954, when one was caught off Japan. Only then did we learn such basic details as what color this dolphin was (dark brown to black on the back and somewhat lighter on the sides, with a gray or white patch between the flippers).

The gray Atlantic hump-backed dolphin's story is similar. It was described from skeletal remains in 1892, but not seen in the flesh until a specimen was caught off Senegal in 1956. The name comes from a ridge which rises from the back around the dorsal fin. Other than the hump and some white or light-gray markings, it looks much like the bottle-nosed dolphin we're used to seeing perform at Sea World. Its rediscovery is one more reminder of how much we still have to learn about these enigmatic mammals.

The Santo Mountain Starling

Most birds are well-known creatures, being by nature mobile and highly conspicuous. However, a few have chosen remote habitats which make them challenging subjects to study.

The Santo mountain starling is one such recluse. It exists only on the Pacific island of Espirito Santo, where it keeps to altitudes above 4,000 feet. Accordingly, the starling was not discovered until 1934, and then not seen again until 1961.

Another thirty years then passed without confirmation of the bird's continued existence. At least two expeditions had looked for the bird in vain before 1991. In that year, another scientific team climbed Peak Santo in a last effort to rediscover the missing starling. They succeeded. In fact, the birds proved to be in decent health—the team logged nine observations in five days—and appeared rather curious about the intruders into their rarely visited neighborhood.

The team caught one starling, which was studied and released. The expedition also took the first photographs ever of this sparrow-sized bird, which is decked out mostly in dark, mottled plumage with a rust-colored breast. The current population appears small but stable.

Even birds on more populated islands can be difficult to locate. Hawaii's resident honeycreepers include the 'Akiaola, a little bird with an absurdly long bill. The forests of Kauai may or may not still conceal this species, which has not been definitively reported since 1973. The same island houses the Kauai 'O'o, already thought extinct until it was rediscovered in 1960. A search in 1981 found only one pair, and no one knows whether the species still exists today.

Maui's nukupu'u, another honeycreeper with a downcurved bill, was luckier. After seventy years of presumed extinction, it was rediscovered in 1967 in the Kipahulu Valley. A 1993 estimate put the population at thirty individuals. The bird's status is extremely precarious, but at least we know it's still with us.

The Pigeon and the Parakeet

We've already mentioned America's most famous extinct birds, the passenger pigeon and the Carolina parakeet. The last known specimen of each died during World War I in the Cincinnati Zoo. But were these birds truly the last of their kind?

The passenger pigeon was once the most numerous bird in the world. A slightly built pigeon with a long tail, it was covered in bluish gray plumage except for a pinkish underside. The population two hundred years ago is estimated to have numbered in the billions. Early naturalists recorded flocks that took hours to pass overhead.

It wasn't surprising that such an abundant bird should have become a food source. Hunters with guns and nets began carting pigeons to city markets in wagonloads. Sport hunters shot millions more, which were usually left where they fell. Destruction of the forest habitat along their migration route was another cause (some experts say the main one) in the birds' drastic decline. Incredibly, the species was reduced by 1914 to the one bird living out her days in a cage. "Martha" died on September 1 of that year, and her mounted body now rests in the Smithsonian.

In 1929, Professor Philip Hadley documented several sighting reports and personally got a brief glimpse of a passenger pigeon in northern Michigan. Sightings have trickled in ever since, with two reports made as late as 1965.

British zoologist Dr. Karl Shuker, who makes a specialty of the odd corners of the animal world, doubts the pigeon is still with us. He points out that its mating behavior involved meeting in huge flocks, and believes that stragglers—of which he agrees there were a few—could not have changed their lifestyle drastically enough and quickly enough to ward off extinction.

The Carolina parakeet, a handsome green bird with a distinctive yellow head, was North America's only native parrot. It was rarely killed for food, but was destroyed in large numbers by orchard growers, who considered the fruit-eater a pest. Others were taken for pets, shot for sport, or harvested to make ornaments for women's hats. The last known wild specimen was collected in 1901. "Incas," a male, passed away in his cage on February 21, 1918.

The parakeet has a stronger case than the pigeon for continued existence. In 1926, one Charles Doe, Curator of Birds for a Florida university, reported finding three pairs in Okeechobee County. This man of science promptly collected their eggs as specimens!

In 1936, ornithologists Alexander Sprunt and Robert P. Allen spotted a flock of probable Carolina parakeets in South Carolina's Santee Swamp. The following year, Sprunt, along with ornithologist Roger Tory Peterson and National Audubon Society president John Baker, reported seeing a single immature parakeet in the same location.

The report of such august witnesses didn't prevent the swamp from being converted into the Santee-Cooper Hydroelectric Project, but it does stand as credible evidence for the bird's survival nearly twenty years after its supposed extinction. Could other swamps still hold Carolina parakeets today? The odds are not good, but there remains at least a shadow of a chance.

Any new finds, unfortunately, cannot be compared with the remains of the lonely Cincinnati Zoo bird. That precious specimen was ticketed for the Smithsonian but never arrived, and no one knows whether it still exists. It thus represents its species all too well, providing the final tragic irony of the Carolina parakeet story.

Bringing Back the Quagga

The quagga was an animal that looked rather like the Creator's rough draft for the zebra. This inhabitant of the South African savanna had brown stripes which covered only the head, neck, and the foremost portion of the back. The rest of its back was a lighter brown, and the underside and legs were cream-colored.

The quagga's closest living relative is Burchell's zebra, also known as the plains zebra. Burchell's zebras are normally striped all over the body, but an occasional specimen turns up in which the stripes on the hindquarters are so faint they're almost invisible.

The quagga suffered from hunting and from competition as domestic livestock was introduced. The last captive specimen died in a zoo in Amsterdam in 1883. The exact date of the quagga's passing in the wild is uncertain, but was presumably around the same time.

There have been occasional reports of surviving quaggas, but no proof, so these may concern the incompletely striped Burchell's zebras previously mentioned.

Building on the relationship between the quagga and Burchell's zebra (recent genetic tests identify the former as a subspecies of the latter), researchers in Africa have chosen zebras with incomplete stripes or brownish hides and attempted to recreate the quagga. Since 1988, this effort has produced ten foals. One born in November 1991 showed both the proper coloring and the quagga's striping pattern, and the encouraged researchers have continued their work, hoping to use embryo transplants and artificial insemination to speed the process of breeding "pure" quaggas.

The experiment is still incomplete, but the long-lost quagga—or at least a living model of it—may yet graze the African plains once again.

Looking for Lemurs

Madagascar, like Australia, is something of a zoological attic, filled with animals the rest of nature forgot about or never met. The world's fourth-largest island drifted away from the African continent over a hundred million years ago, and its animals have developed in isolation.

The dominant mammals on Madagascar are the small primates (at least, the living species are small) known as lemurs. These animals seem to have a talent for toying with inquisitive zoologists.

Take, for example, the rat-sized species called the hairy-eared dwarf lemur. The first specimen of this gray-brown animal, sporting large eyes and tufted ears, was found in 1874: the second, not until 1965. No more turned up until 1990. In that year, Bernhard Meier, a zoologist from Germany's Ruhr University, went searching for the possibly extinct animal. He found it—or, rather, it found him, obligingly leaping into the beam of his headlamp. He observed a total of three animals, one of which he caught, examined, and released. Two pairs have since been captured for study and possible breeding.

The lemurs today number twenty-eight species, not counting the fourteen known extinct types. For a long time, no one was certain which category the greater bamboo lemur belonged in. One of only two known lemurs—and a handful of mammals in general—that can live on that giant grass, it was discovered in the late 1800s and then vanished into the forest. An olive gray animal weighing about five pounds, it was classed as an extinct species until 1964.

In that year, a French naturalist named Andre Peyrieras rediscovered the animal when he encountered a specimen for sale in a rural marketplace. Despite intense scientific interest, the lemur was not seen again until Peyrieras and three colleagues spotted another one in 1972.

Its status remained uncertain until an international team set out in 1986 hoping to find the animal once more. Duke University's Patricia Wright did indeed spot a lemur eating bamboo. But the color—a distinctive russet—was wrong, and the golden eyebrows and orange whiskers were *very* wrong!

The expedition had found a new species, the golden bamboo lemur. On an island that had already lost the vast majority of its forest cover, it was a thrilling find.

Just to put the icing on the cake, the same expedition redis-covered the greater bamboo lemur as well. Wright counted thirty-six of them, to be exact. No zoologist could ask for a better trip.

Efforts are underway to set aside reserves for lemurs, as well as to breed some species in captivity. The survival of many lemurs, however, remains in doubt, and rare species like the greater bamboo lemur may vanish again into uncertainty. It's all up to their distant relatives, the humans.

The Aye-Aye

None of Madagascar's lemur family members looks stranger than the nocturnal aye-aye, with its black fur and huge pale eyes. The clawed third finger on each hand is disproportionately long and thin, forming a specialized tool for digging grubs from rotting wood. Such grubs, detected by the primate's oversized ears as they move beneath the bark of trees, form a major part of the aye-aye's diet. From an ecological point of view, the aye-aye is a sort of mammalian wood-pecker, filling the niche left empty by that bird's absence from Madagascar.

This peculiar primate looks like some sort of goblin. Local folklore treats it that way, considering the aye-aye a harbinger of death that points its strange middle finger at victims to seal their fate. The aye-aye, sometimes called the witch-cat, is a creature to be avoided—and, in some areas, killed instantly if it *is* encountered.

Between development and superstition, it's not surprising the aye-aye's population shrank. Accordingly, it was thought to be possibly extinct until two adults and a juvenile were captured in 1986. These became the first aye-ayes in captivity. In 1992, the first successful captive birth was announced at Duke University, although this baby had been conceived in the wild. The little goblin was promptly christened Blue Devil after the school's mascot. Shortly afterward, British scientists at the Jersey Wildlife Preservation Trust reported the birth of the first aye-aye actually conceived in captivity. As Madagascar's impoverished population continues destroying the forest to scratch out a living, the aye-aye will need all the help it can get.

Tigers in Twilight

"Tiger, tiger, burning bright/in the forests of the night," wrote William Blake, marveling at the animal's "fearful symmetry." Sport hunters, however, have never had the same reverence for the animal. Along with fur hunters and the pressures of human development, they have reduced several subspecies to the point where we don't know whether the tigers still lurk in the forests. Any that remain are protected by the Convention on International Trade in Endangered Species of Wild Fauna and Flora, known as CITES, which bans any trade in products of the most threatened species.

The smallest member of the tiger family, the light-colored Bali tiger, apparently became extinct around 1937, leaving only rumors and shadows of sightings. Swedish zoologist Kai Curry-Lindahl believed that three or four were still alive in 1963, but he visited the island two years later and failed to find any tigers. Fresh clawmarks found on trees in 1979 gave rise to a glimmer of hope, although it's hard to believe enough tigers remain to make up a viable population. (The World Wildlife Fund researchers who documented those signs also found clawmarks and a track apparently left by a leopard, an animal never known to live on Bali.)

The Javan tiger, distinguished by thin stripes and dark red fur, may have met the same fate. As the population dwindled due to shooting and habitat destruction, experts debated whether to capture the last few for captive breeding, but the debate dragged on until there was apparently nothing left to breed.

A study in 1976 counted five living individuals. John Seidenstecker found tracks of three in 1979. They may all be gone now: any that remain continue living out their normal tiger-lives, but they are the walking dead of their kind. In fact, the IUCN sadly wrote the subspecies off in 1984. Still, tiger sign was discovered in 1990 and again in 1992. In November of 1992, a Javan newspaper reported that a tiger had been photographed. The International Society for Endangered Cats funded a camera-trap project starting in 1993 to investigate a continuing trickle of sightings.

The larger Caspian tiger may also be extinct. It once ranged across much of central Asia, but no recent sightings have been confirmed in Turkey or Russia. A pawprint cast taken in Iran almost twenty years ago recorded the passage of a solitary animal, perhaps the last one in that country. Today, a few may linger in the mountains of Afghanistan.

The South China and Sumatran tigers definitely still exist, but are dwindling rapidly. There may be as few as thirty South China tigers left. The enormous Siberian tiger, the most majestic and perhaps the original member of its family, numbers only a few hundred in the wild and suffers from widespread poaching, although it at least has a healthy captive population.

Peter Jackson, chairman of the IUCN's Cat Specialist Group, reports grimly that, "My belief is that the end of the tiger is in sight, possibly within ten years."

The snow leopard is in a similar predicament. This reclusive Asian mountain-dweller, the most beautiful and the least known of the great cats, was never even photographed in the wild until 1970. Demand for its luxuriant smoky gray coat, marked with gray-black rosettes, is driving the animal to extinction. Despite a worldwide ban on sales of snow leopard skins, the unconscionable trade continues, and the snow leopards in zoos may soon outnumber their wild relations.

That zoo population ensures the leopard's survival, but an animal that lives on only in captivity is not quite the same: in some indefinable way, the species has lost part of its collective soul.

The Black-Footed Ferret

There aren't many cuter animals in North America than the black-footed ferret. A nocturnal predator about eighteen inches long, this little relative of the weasel is mostly brown, with round ears and a black mask around its eyes.

Cuteness was no help, however, when humans began eradicating the "nuisance" prairie dog, the ferret's main prey animal. Those efforts led to the ferret's presumed extinction by the 1950s.

In 1964, a ferret sighting in Mellette County, Wyoming, led to a study that counted twenty-one of the animals. Ten years later, the ferret had vanished again, apparently for good.

In 1981, the ferret appeared in Wyoming once more. It was rediscovered, not by scientists, but by a mixed-breed dog named Shep. Shep, not being interested in live specimens, killed his discovery, but investigators found that Shep's dinner was not the only ferret left.

This second resurrection of the ferret set off something of a scientific carnival, as federal, state, and private researchers swarmed into the area. Every ferret in sight was trapped, radio-collared, studied, and ear-tagged, sometimes until their ears became tattered.

The ferrets were decimated a few years later by distemper (aggravated, some charged, by the stress of being studied half to death). Biologists captured all the remaining animals they could find, a total of eighteen, for a captive breeding program.

This last-chance approach is always controversial. Some conservationists see an animal's removal from the wild as a tragically necessary step, while others denounce it as a kind of sacrilege. Almost everyone charged that such measures wouldn't have been necessary if the ferrets had been better managed all along.

Whatever the truth in these arguments, the program has saved the ferret. Today there are more than 300 animals in captive colonies. In 1991, the first carefully monitored release into the wild took place when forty-nine ferrets were returned to the Shirley Basin in Wyoming. Despite a high mortality rate (not unknown in reintroductions), six ferret youngsters were spotted in 1992.

More release efforts followed, and 187 animals had been turned loose by 1994. Captive breeding is continuing, and animals slated for release are now sent to "ferret basic training" at a former Army depot in Pueblo, Colorado. There, in storage buildings converted into prarie dog towns, the ferrets learn to hunt on their own. The species still has a long way to go, but the ghostly little "outlaw" of the prairie is finally back where it belongs.

The Mysterious Starling

This story begins with a lone specimen resting in the vast collections of the British Museum. It was about seven inches long, almost entirely brown, save for a slight bronze tint on the head and a little white on the breast. Its label read "The Mysterious Starling."

The name certainly fit. The bird had a scientific name and an officially published description, but no one knew where it came from (the Pacific was supposed) or when it had been collected.

Storrs Olson of the Smithsonian Institution apparently solved the mystery in 1986. In an old manuscript also held by the British Museum, he found an account by a ship's naturalist named Andrew Bloxham. Bloxham had been on board the HMS *Blonde* in 1825 when it delivered the bodies of the king and queen of Hawaii to their home islands for burial (they had been visiting England, where they died of measles).

On its homeward voyage, the *Blonde* made a brief stop at Mauke in the Cook Islands. There Bloxham took a specimen of a local starling, whose description matches the one in the museum today.

The bird, however, is still mysterious enough. In the more than one and a half centuries since its collection, no one has ever seen it.

The Cahow

The record for the longest apparent extinction in historical times may be held by the cahow, or Bermuda petrel. This handsome black-and-white seabird, with a wingspan of three feet, was once common enough to rate a mention in Shakespeare's *The Tempest*. By 1616, however, its numbers had been so reduced by hungry settlers and their animals that the bird was the subject of the New World's first game law, a proclamation from the Governor forbidding the "spoyle and havock" of cahows. Despite this, the species was believed wiped out by 1621.

After that, the cahow was almost forgotten. No specimens or skeletons were preserved, and no one had even recorded a proper description for posterity.

In 1906, Bermuda Aquarium director Louis Mowbray caught a seabird he believed was a New Zealand petrel. It was stuffed and added to the aquarium's collection. Ten years later, cahow bones were discovered in island caves. Comparing them to his "New Zealand petrel," Mowbray realized he had made a momentous discovery without knowing it. He had found a cahow.

Still, one specimen meant only that a single cahow had been alive in 1906. It was not proof the species was still viable. In 1935, the count became two cahows when a second specimen flew into the window of a lighthouse.

This bird was identified by Dr. Robert Murphy of the American Museum of Natural History. In 1941, a cahow flew into a telephone cable, and four years later an ornithologist serving as an American military officer found a cahow carcass on Cooper's Island.

In 1951, Dr. Murphy finally took an expedition to Bermuda to determine the cahow's status once and for all. In January of that year,

Murphy and his assistants discovered seventeen cahow nests in burrows scattered among five islets near Castle Harbor. Seven adult pairs were counted. With careful protection, these increased to thirty-five by 1985. Young cahows spread out to new nesting sites in the area, helped by conservationists' success in persuading a local U.S. military installation to dim floodlights which interfered with the birds' mating habits.

The cahow remains endangered. Breeding has been slow, a circumstance some blame on DDT poisoning. Another predatory seabird, the white-tailed tropicbird, killed some young cahows in their burrows until baffles were fitted over the entrances. These were designed to let the cahows through but keep out the larger tropicbirds.

Finally, like any localized population, the cahow could easily fall victim to a hurricane or epidemic. But after a species has been missing for three hundred years, any chance of survival is an improvement.

Birds carrying the name "petrel" seem to share this penchant for disappearances. Consider the Fiji, or MacGillivray's, petrel. After the type specimen was obtained in 1855 by the HMS *Herald* on the Pacific island of Gau (or Ngau), this small black seabird dropped out of sight and out of the ornithological reference books. It existed only as a footnote, presumed to be a freak, an extinct form, or both. There was even uncertainty about why John MacGillivray's name was attached to the find. He *was* the *Herald*'s naturalist, but was not on board when the ship visited Gau. He was busy facing a court of enquiry in Australia, where he was convicted of selling ornithological specimens belonging to the government.

Subsequent searches failed to locate the bird, but British naturalist Dick Watling, a Fiji resident, never gave up. In 1984, he tried using lights to lure birds in from the sea at night. The result was that a live Fiji petrel was rediscovered when it crash-landed on the researcher's head!

A third petrel that played this game was the taiko of the Chatham Islands, which was known only from a type specimen found at sea in 1867. The species emerged from limbo in 1978, when two birds were captured alive. After study, the rare birds were released, hopefully to continue their shy species' existence.

The Pygmy Hog

The pygmy hog looks like a miniature version of most of the world's other wild pigs. The head and body length is only about two feet, and a full-grown adult may be less than a foot tall. The hog's hindquarters stand higher than the shoulders, tilting the animal permanently forward. Its bristly fur is mostly dark brown, sometimes with a reddish cast.

The hog's natural habitat is in the southern foothills of the Himalayas, mainly in India. There the hog digs a trench, then bends the long grass over it, creating an unusual nest which resembles a miniature hut.

Once this homestead is established, the hog spends almost all its time foraging. Like most hogs, it eats virtually anything. Unfortunately, another trait the pygmy shares with fellow swine is that it makes tasty pork chops. Heavily hunted, the hog faded out of sight in the mid-twentieth century and was generally believed to be gone for good.

The species reappeared in dramatic fashion in 1971. A series of destructive fires swept through the tall grass of its habitat and flushed out live pygmy hogs, some of which were captured and taken to zoos. (Others, unfortunately, were captured and taken to kitchens.)

Today, small numbers are kept in captivity in India, although an attempt to breed some in Zurich failed. The hog has been protected by Indian law since 1972, but it remains so rare that it made the IUCN's list of the world's twelve most endangered animals in 1984. Its habitat continues to shrink under the pressure of agricultural development.

An interesting sidelight is that those devastating 1971 fires also led to the rediscovery of a second presumed-extinct mammal. This was the hispid hare, a large member of the rabbit clan not seen since 1956.

Bachman's Warbler

The World Wildlife Fund lists Bachman's warbler as "the rarest native songbird of the U.S."

It's a small bird, less than sparrow-sized. The male's plumage is olive on the back, with a yellow face and underside and black "cap" and "neckerchief" markings. As with most birds, the female's colors are less vibrant. The male's song is memorable, described by one author as being performed with "a passion and intensity that impressed itself on all who heard."

No one knows for certain whether Bachman's warbler still exists. The species breeds—or bred—in the forested bottomlands of the Southeast, spending the winters in Cuba. Its numbers suffered at the hands of feather hunters and developers in the U.S., but the heaviest blow came from the wholesale clearing of Cuban forests to plant sugarcane. Pairs do not winter together, and, as the total number of warblers declined, birds returning from Cuba faced greater and greater odds against running into a mate in the much larger U.S. breeding territory.

Bachman's warbler was never a common species, and by 1967 it had to be labeled "endangered." This may have been an understatement, because the last known active nest was reported from Alabama in 1937. All subsequent sightings involved lone birds. By the mid-1970s, these had dwindled to occasional reports, mainly from South Carolina. Since a female was spotted in Cuba in 1981, the species may not have been seen at all.

The current state of Bachman's warbler is "unknown, possibly extinct." The little songbird may yet be holding out somewhere in its range, with the swamps of South Carolina's Francis Marion National Forest considered the most likely location. American birders are keeping their binoculars polished and their fingers crossed.

You might think a bird couldn't be long overlooked in North America, given the continent's large population and enormous number of birders. Consider, however, the case of the giant Canadian goose. *Branta canadensis maxima* was only established as a distinct subspecies in 1951. By the time of that recognition, unfortunately, the giant (one adult weighed twenty-four pounds, remarkable for any flying bird) was already presumed extinct.

In 1960, the subspecies was rediscovered in Canada. Two years later, a completely overlooked breeding population was spotted in Minnesota. If one of the largest flyers on the continent could go unnoticed, we shouldn't be too quick to write off a diminutive warbler.

Bandicoots and Wallaroos

In the previous section, we met some of the newest marsupials. After so many remained unknown for so long, it shouldn't be surprising that the status of other species is highly uncertain.

From the Indonesian island of Seram comes the long-nosed bandicoot, another animal with one of those whimsical names suggesting that the original marsupial taxonomist may have been Winnie-the-Pooh. It looks sort of like a big mouse, about eighteen inches long with the tail.

At least, this dark brown marsupial with a sparsely furred tail *did* come from Seram. Four specimens were collected in 1920 from a dense mountain forest. The animal has not been seen since. The New Guinean mouse bandicoot, described in 1932, is also known from only four examples.

The rufous bandicoot of Australia was also discovered in 1932. About twenty inches long, it has bristly fur and an unusually fragile tail which is often broken or lost. A resident of the Cape York

Peninsula, it disappeared after its discovery and wasn't found until the 1970s. It now turns out this animal's population is substantial.

Leadbeater's possum is a handsome gray-furred animal with black and white markings, about a foot long overall. It was discovered in 1867, but dropped out of sight after 1909. Only five specimens had been collected. The animal was rediscovered in 1961 in a forest in Victoria, Australia.

The desert rat-kangaroo, with a total length of about thirty inches, is another "disappearing" animal. Rat-kangaroos look more like oversized rats than kangaroos. Pale yellow-brown above and white below, this species was found in 1843 and then went missing until 1931, when new specimens were found in Australia's Lake Eyre Basin. The last sighting, though, was in 1935. It may be extinct, or it may just be pulling its vanishing trick again. The same era brought the first and last encounter with the central hare-wallaby, a mysterious species known from one skull.

The bridled nail-tailed wallaby, a ten-pound member of another mini-kangaroo family, is a brownish animal with a broadening white stripe that curves attractively from the head down the flanks. The "nail-tailed" appellation derives from the tip of the tail, which forms a hard, pointed spike of no apparent use. The wallaby was once abundant and wide-ranging in Australia, but land clearing and introduced predators apparently wiped it out some sixty years ago. It stayed "wiped out" until 1973, when a Queensland contractor recognized one from a picture he'd seen in a magazine. Its relative, the crescent nail-tailed wallaby, was last spotted alive in 1956 (a carcass was found in 1964) and remains in the maybe-extinct twilight zone to this day.

The black wallaroo, yet another kangaroo offshoot, is a good-sized animal, reaching over forty pounds. A resident of Australia's Northern Territory, it was last seen in 1914 and was almost forgotten by the time it bounded into view again in 1969.

The parma, or white-throated wallaby, has an even more interesting story. Smaller than the wallaroo, it is brown with a white underside. It was believed extinct in its habitat of New South Wales in the early 1930s.

It was rediscovered in 1965 a long way away: on the island of Kawau, off New Zealand. It turned out some of the animals had been transplanted to the island and released on a private estate before the turn of the century. This was unknown to zoologists, who were giving the animal up for lost while farmers on Kawau were shooting the abundant "pests."

That wasn't the only reappearance trick the parma had up its sleeve. Two years after the rediscovery on Kawau, it was found again in its original location on the mainland.

The freckled marsupial mouse is also known as the dibbler. Its range in South and Western Australia is quite limited today. In fact, the animal apparently disappeared before the turn of the century. In one of those startling reappearances Australian mammals seem to specialize in, it was found again in 1967.

Most recently, in December, 1994, the rediscovery of Gilbert's potoroo was announced. Potoroos look a bit like chubby rats with kangaroo hind legs, and this particular type hadn't been seen in the twentieth century and was flatly listed in most reference books as "extinct." That's a word Australian zoologists must be getting very hesitant to use.

Wanted Birds

Bird species have appeared and disappeared throughout ornithological history. A surprising number have been seen only once. Such cases present questions of existence that frustrate researchers for years, decades, or even centuries. Let's look at just a few of the mysteries the bird world has presented.

A subspecies of the Nightingale reed warbler is known from two specimens collected at an unrecorded Pacific location by a French expedition in the 1830s. This bird, *Acrocephalus luscinia astrolabii,* has not been located since.

The Ryukyu kingfisher is about nine and a half inches long, with a head and underside mainly feathered in cinnamon and a blue-green back. The wings are brownish black, the beak long and sharp. The only known specimen was caught in 1887, and now rests in a Tokyo museum. It's not even certain it came from the Ryukyu Islands in the first place. Wherever the bird originated, it has been hidden—or extinct—for over a hundred years.

In the 1780s, naturalist George Forster accompanied Captain James Cook on his second voyage around the world. On that expedition, Forster described the Tahitian red-billed rail and painted its portrait. That painting is in the British Museum, but, while Forster's account is generally accepted, this small black and white rail with a red beak and pink legs has not been rediscovered. Native accounts of the bird, which they called tevea, oomnaa, or eboonaa, indicate it may have existed on the island of Mehetia until the twentieth century.

One example of the dove *Gallicolumba ferruginea* was caught on Tanna in Vanuatu in 1774. That bird, which sported a brown back and red-orange breast, remains the only known specimen. That same year, Captain Cook's expedition collected two examples of the Society Island parrot. It, too, was never seen again.

The New Caledonian owlet nightjar, a dark, mottled bird, is the largest member of its family. The only preserved specimen was captured in 1880 when it flew into someone's bedroom. The bird hasn't been caught again, although a hunter reported shooting one in 1970.

Also from New Caledonia came Lafresnaye's rail, an olive-brown bird which lived in the swamps of Dumbea and Saint-Louis. None has been collected in fifty years, but it may still be alive. Author Jean-Christophe Balouet reports a possible sighting in November 1984.

The white-winged guan, a member of the gallinaceous family which includes the pheasants and quails, was described from two nineteenth-century skins collected in northern Peru. It was believed extinct until its rediscovery in September 1977, almost exactly 100 years after the bird was first described. On the other side of the world, the Madagascar serpent eagle, missing since 1930, was seen alive by ornithologists in 1988.

Charles Darwin's Galapagos finches are famous for the role they played in his concept of natural selection. Among the many he collected in 1835 are seven examples of a species called Darwin's large ground finch. While smaller relatives of this bird are common, the much larger species was not seen again until 1957, when a single bird was collected on Floreana. No more have turned up since.

As yet another reminder of birds' ability to remain overlooked, consider the Mongolian, or relict, gull. Known from a type specimen collected in China in the 1930s, this bird's continued existence and even its status as a species were in dispute until a nestling was found and banded in 1974. Breeding populations are now known to have survived in Vietnam, China, and Kazakhstan.

Finally, we have the uncertain fate of the world's smallest ostrich. The Arabian, or Syrian, ostrich is probably but not definitely extinct. It was common until after World War One, when the widespread introduction of firearms hastened its decline at the hands of local hunters. One account is that an ostrich shot in Bahrain in 1941 was the last. Another report is that a German tank crew may have killed and eaten the last one around 1942. However, a flood in southwestern Jordan in February 1966 carried with it a dying ostrich. Do any more still roam the deserts of North Africa or the Middle East?

The Thin-Spined Porcupine and Company

Brazil's thin-spined porcupine is an odd little mammal that seems to belong somewhere between the true porcupines and a family called the spiny rats. The pudgy rodent is about eighteen inches long, including a unique tail covered with bristles on the underside and scales on top. The porcupine was photographed once in 1952, then lost for more than thrity years and considered extinct before a local hunter led a Brazilian biologist to one of the creatures.

The three-banded armadillo pulled a similar stunt, being classified as extinct until 1991. Brazilian graduate student Ilmar Bastos

Santos was the rediscoverer. He found the animal in, of all places, a local market, where it was being offered for sale. Its habitat is in northeastern Brazil, where it is known as *tatu bola*. The armadillo apparently has been hiding out in dry regions called caatingas. These areas also house such rarities as Spix's macaw, a bird whose known wild population totals exactly one animal.

Looking at problematical small mammals in general, you'd expect mice to be good at this sort of hide-and-seek. The record agrees. The Hastings River mouse of Australia vanished around the 1840s and stayed vanished until 1969, when it was rediscovered in Queensland. Another Australian mouse, known as the Heath rat, was "lost" in Western Australia in 1906. It was found again in another province, Victoria, in 1961.

The insectivores are also known for disappearances. One family, the tenrecs, live mostly on Madagascar. Several species are known from one or a few examples only: *Microgale pusilla,* from the Ikongo Forest, has not been seen in nearly a hundred years.

The golden moles are equally secretive. These are tailless, sometimes eyeless inhabitants of Africa. Some are truly golden, while others are brown or gray. Visagie's Cape golden mole, from South Africa, was found once in 1950. Van Zyl's golden mole, from the same area, is also known from just one example, described in 1938. The Somali golden mole's status is the same: its lone representative was described in 1968.

Shrews are small even by insectivore status. They are voracious animals that humans find thoroughly unpleasant. *Blarina carolinensis shermani,* a subspecies of America's southern short-tailed shrew, is one of the rarest, if it's around at all. It is known only from a small area of Florida.

Finally, not all rediscovered mammals are all that small. Pakistan's woolly flying squirrel, the largest squirrel in the world, was discovered in 1888, vanished soon after, and stayed missing until 1995. The animal, which is up to two feet long not counting its two-foot tail, was found by two dogged amateurs after eluding repeated searches by professional zoologists.

For mammologists who'd like to make the record books as rediscoverers of a lost species, there are plenty of opportunities.

Shadowy Bats

Bats, even more than most mammals, can be difficult to keep track of. Unless they can be tailed to their roosts, these little riders of the night air make elusive targets for study.

The Halcon fruit bat was discovered in 1937, on the slopes of Mount Halcon on the Philippine island of Mindoro. This brown-furred bat has not been seen since. The bare-backed fruit bat, a reddish brown bat about four inches long from Indonesia, is in the same state. Specimens found in the 1930s are the only examples to date.

Also from Indonesia comes the Sulawesi dawn bat, known from only one specimen. Ridley's leaf-nosed bat, a diminutive creature from Malaysia, was "lost" in 1910 and not rediscovered until 1975.

The red fruit bat had a stranger story. It was discovered back in 1820 on a Carribean island—which one is not certain. The first specimen remained the only one known until 1916, when it was rediscovered on Puerto Rico—as a fossil! A small number of live specimens have since turned up on Puerto Rico and in the Virgin Islands.

The Tanzanian woolly bat was discovered in the 1870s and has not been seen again. The big-eared bat (another tiny one, despite the name) has been missing since its Papua New Guinea discovery in 1914. The large-eared free-tailed bat was known only from the type specimen found in Kenya until two were found in the 1970s. However, the new examples were found 1500 miles away in Zimbabwe. The intriguingly named Salim Ali's fruit bat was discovered in India in 1948, then immediately dropped from sight and was not found again until 1993.

The fossil fruit bat's name is a fitting one. The bat was known due to fossil remains, at least nine thousand years old, from sites of human habitation. Did ancient hunters wipe the bat out? It was thought so until 1975, when a single living specimen was procured from the mountains of the Hindenburg Ranges in New Guinea.

That specimen has since been lost, leaving only the fossils—until and unless the bat surfaces once more.

Numerous other bats are known from one or only a few specimens: there's no point in going through the whole list. The point is that we know darn little about these mammals. As mentioned, they are hard to find: besides, not many people want to know anything about them. There is plenty of room left for bat research.

The Lonely Tortoise

The tortoises of the Galapagos Islands are world-famous. These waddling fortresses thrived for millenia before humans came along. Today, as a result of introduced predators and indiscriminate hunting to stock the larders of passing ships, several subspecies are extinct.

The saddle-backed tortoise of Pinta Island, a thin-shelled reptile about three feet long, was classified as one of these lost forms until 1972. Then a lone survivor was discovered. George, as he is known, is living out his kind's last days at a research station on Santa Cruz. He is reportedly a shy old fellow who hides from humans, for which he can hardly be blamed.

Attempts were made to match George with two females of a closely related variety. One he ignored: the other he drowned. That was the end of George's love life.

In Bangladesh lives a black soft-shelled turtle that definitely exists—anyone can visit its habitat—but whose origin is a mystery. *Aspideretes nigricans* has a surviving population of about 350. All live in the same place: an artificial pond at a shrine in Chittagong. The captives, some of whom weigh sixty pounds, live on morsels offered by pilgrims to the shrine. This is the only place these turtles have ever been seen. They have lived here for over a thousand years, and where they came from originally is unknown.

Turning to other members of the reptile clan, lizards offer some puzzles of their own. One is a monitor lizard discovered as a fossil in 1983 on New Caledonia. Jean-Christophe Balouet, author of *Extinct Species of the World,* reports he talked to several people who said they'd seen this "fossil" alive on the island's northeastern coast. One person claimed to have killed one about 1970.

On the Cape Verde Islands lived the largest member of the skink family, a mottled brown lizard about two feet long. This reptile may be extinct: the last definite reports date from the 1940s. Then again, it may just be hiding out in the forest.

That was the case with the Jamaican iguana. This lizard, the island's largest endemic land animal, was believed extinct in 1946 after introduced mongooses apparently wiped out the population. There were occasional sighting reports from the rugged forests of the Hellshire Hills, but these went unconfirmed until 1990, when a hunting dog brought its master a live iguana. Frantic searches produced evidence of a surviving population numbering a hundred or less. Small groups have now been relocated to two U.S. zoos for captive breeding, and the Hills have become a protected area. The fate of what may be the world's rarest lizard still hangs in the balance.

Jerdon's Courser

Nigel Collar of the International Council for Bird Preservation (ICBP) called it the "bird conservation high spot" of the 1980s. The event: the surprising reappearance of an almost unquestionably extinct bird, Jerdon's courser.

The subject of all this excitement is a ground-dwelling bird of India's state of Andhra Pradesh. It's a long-legged, plover-like creature measuring about ten inches long. The plumage is mostly sandy brown, with a white underside and decorative white bands reaching around the head and across the breast.

Jerdon's courser was always rare. Discovered in 1848, it was seen only three times before 1900. Then there were no more sightings. From the 1930s on, four searches were mounted to rediscover the bird. All failed to find any trace of it. Adding their failure to the fact that almost ninety years had gone by without a sighting, it's hard to say how the bird's extinction could have been more certain.

By the 1980s, only the Bombay Natural History Society refused to give up. In a last effort, ornithologists roamed the area with color posters of the bird and heard several recent sighting reports. These indicated the courser was nocturnal, and searches were made by flashlight.

In January 1986, a birder caught a live Jerdon's courser in his flashlight beam. The bird was dazzled by the light, and the hunter simply walked up and grabbed it.

A still-unresolved avian puzzle from India involves the pink-headed duck, a colorful fowl named for its striking pink head and neck. Threatened by hunting and by the draining of its favorite swamps, the duck may have disappeared as early as 1935, when the last reported wild specimen was shot. The ICBP classifies it as extinct, effective that year.

As has happened in so many other cases, this may not be the last word on the subject. A game warden named K. L. Mehta reported seeing a pink-headed duck in Kunihar State in 1960. After his report was publicized, a story came out concerning a shooting in 1947 and other sightings in 1948 and 1949. In 1974, an article by Richard Fitter of the International Union for the Conservation of Nature noted a "recent plausible sighting" in Nepal.

In 1987, adventurer Rory Nugent made a difficult journey through northern India in search of the duck. "Difficult" is an understatement: Nugent faced brushes with arrest, gunfire, and drowning. He also had to fend off innumerable Indians trying to sell him unfortunate ducks with freshly painted pink heads.

Nugent found a few people who claimed to remember sightings, but he never spotted the duck. He did confirm that large areas of duck habitat along Indian's northern borders were inaccessible for reasons of security, red tape, or rough terrain, making a definitive search almost impossible. The pink-headed duck's case is not yet closed.

Section III

The Mystery Animals

Introduction: The Continuing Search

No one doubts there are still animals to be found. Even today, there are regions of Africa, South America, and Asia that remain largely unpopulated and unexplored. Other areas are inhabited only by indigenous cultures whose members have no reason to send descriptions of the local animals to an outside world they may be doing their best to avoid.

It is the large and unusual animals, of course, that get the most attention, although their stories may be the least well documented. A lumberman who thinks he saw a North American ape will make the papers: a scientist who believes she spotted a small wild dog in Bolivia draws considerably less attention, but the latter is more likely to have made a real discovery.

The real discoveries are what concern us here. The animals described in this section are those we have solid reasons to believe in.

In some cases, such as the pygmy elephant, zoos have held living specimens: the problem is in deciding whether these animals represent a bona fide species or are merely freak individuals. With other animals, such as MacFarlane's bear, a skin, skull, or other hard evidence remains. Here again we have the "freak" question, plus the possibility that the evidence came from a species which has since faded into extinction.

Finally, there are cases where there is no hard evidence, but a qualified scientist has described the animal from close range. The mammal Georg Wilhelm Steller called the "sea monkey" is a good example: it remains worthy of consideration because the source was a reliable one. Zoologists are reluctant to formally describe a species without an example (called a type-specimen) in hand, but this kind of evidence does mandate keeping an open mind.

Some of the animals in this section probably will prove to be individual mutations or observers' mistakes. Others may remain in limbo forever. But at least some of the animals presented here will one day appear in our textbooks, and perhaps even our zoos. Which ones these will be makes for an endlessly fascinating guessing game for zoologist and lay reader alike.

The Onza

Inhabitants of western Mexico's Sierra Madre Occidental region have been reporting a strange big cat since Aztec times. Records left by Europeans go back to the Spanish conquest, when the invaders saw both a puma and a leaner, unrecognized type of "lion" in Montezuma's royal zoo.

The onza was generally consigned to folklore, despite reports from Americans such as Dale Lee. Lee, a professional hunting guide who had shot nearly 500 panthers, reported killing an onza in 1938. He wrote that the slender, long-eared animal "differs from any of the cat tribe I ever saw." J. B. Tinsley, in his 1987 book *The Puma*, reproduces photographs of this cat and of a similar beast killed by a trapper named Ruggles in 1926.

The onza story languished until 1985, when Mexican rancher Andres Murillo shot a cat he didn't recognize. Fortunately, the man had heard of the legendary onza and was curious enough to contact an interested friend who knew American zoologist J. Richard Greenwell. Greenwell teamed up with a leading mammologist and puma expert, Auburn University's Dr. Troy Best, and the two have been studying the carcass ever since. These researchers have also tracked down two other onza skulls. One of these dates from 1938 (it is not from Lee's specimen, so it seems two onzas were killed that year) and the other from about 1975.

The onza is definitely pantherlike, but has longer legs and a slimmer build. The ears are long, and on the inside of the tawny cat's front legs are unique dark horizontal stripes. The type specimen, a

female about four years old, weighed less than sixty pounds, considerably less than a normal panther its size. The researchers wondered at first whether the specimen was starving or diseased, but it proved to be healthy and well fed.

As Dr. Best explains, one specimen and two skulls don't provide enough material for a conclusion about the onza's classification. The tissue analysis done so far identifies the onza as a panther, but the abnormal build and proportions mark the onza as a "clearly different" animal. We may be dealing with a panther subspecies, a local variation, or even just a recurring abnormality born of normal panther parents. A complication is that there are few "normal" puma specimens from that region of Mexico to compare the onza with.

Whatever the onza finally turns out to be, the most important thing is simply that this "mythical" animal exists. To those scientists with a special interest in searching for new animals, every such find is a welcome reminder that Mother Nature still has her secrets.

Steller's Sea Monkey

Georg Wilhelm Steller, the great early naturalist whom we met in the story of Steller's sea cow, left us a puzzle. Among the numerous Arctic animals and plants he described, all but one have been classified and added to the textbooks. That lone exception is a creature Steller called the "marine ape" or "sea monkey."

The naturalist reported encountering this peculiar swimming mammal on August 10, 1774, while voyaging south of the Alaskan peninsula. He observed it for two hours, sometimes at such close range he "could have touched it with a pole." It was almost five feet long, with fur predominantly the color of "a chestnut cow" and a doglike head with prominent whiskers and erect ears. Steller did not notice any front flippers, but thought he could discern a tail shaped like a shark's as the animal cavorted about his boat in a manner that seemed inquisitive and almost friendly. Sometimes it played with a strange-looking piece of seaweed, which the methodical Steller also described in detail and which is now a well-known species.

The encounter ended when Steller decided to obtain the "very unusual and unknown sea animal" by shooting it. He missed. The creature was seen several more times, but never again close-up.

At a loss to name the beast, Steller called it a "sea monkey" because it reminded him of some descriptions by that name in a bestiary written by Konrad von Gesner two hundred years before. (Zoology texts, reliable or not, were so rare in those days that it wasn't unusual for a work to remain in use for centuries.)

What was this creature? The description, if accurate, rules out not only the known pinnipeds but every known animal, living or extinct. Given that it's hard to believe this eminent naturalist made his story up, we are left with a conundrum.

Steller's major biographer, Leonhard Stejneger, thought the sea-monkey was just a fur seal. He noted that Steller, up to this time, had probably never seen a fur seal. However, as Stejneger's own book recounts, Steller later in this voyage studied, dissected, and wrote in exhaustive and precise detail about fur seals and sea lions, without ever hinting that either one resembled the sea monkey.

One modern biologist, Dr. Roy Mackal, suggests the sea monkey was an immature specimen of an unclassified long-necked, predatory Arctic seal. This hypothetical animal would be a counterpart, perhaps a relative, to the sinuous, aggressive leopard seal of the Antarctic. The leopard seal often tucks its front fins tight against its body when swimming and does display the inquisitive behavior Steller described. The sharklike tail remains a problem, although

seals can hold their rear flippers together in a way that might give the impression of such a tail.

Evidence for an unknown seal's existence in the Arctic is limited to a handful of reports, plus native Inuit traditions. Sadly, Mackal may be right when he suggests that Steller's encounter was the first and last definite sighting of an animal which is now extinct.

Whether this creature is extinct or in hiding, there is no way to dismiss a detailed description by an observer of Steller's prominence. Whatever he saw remains a first-class riddle of zoology.

The Strangest Dolphin

One common feature of whales and dolphins is that the known species have only one dorsal fin. (A few, like the beluga whale and the finless porpoise, get by with none at all.) Like most rules, this one appears to have at least one exception; an animal which has never been caught but whose existence seems fairly well established.

A large dolphin with two dorsal fins has been seen several times in the Pacific and the Mediterranean. The second fin is small and located just behind the head.

A whole school of such dolphins was seen in 1819 off New South Wales by French naturalists Jean Quoy and Joseph Gaimard. These two were well-respected in zoological circles, having also discovered the hourglass dolphin of the same region. The observers wrote that "the volume of the animal was about double that of the ordinary porpoise, and the top of its body, as far as the dorsal fin, was spotted with black and white."

The dolphins swam close by the witnesses' ships, the *Physicienne* and *Uranie,* enabling Quoy and Gaimard to sketch the animals. The drawing they published shows a black body marked with irregular splotches of white. The scientists did not get a good look at the animals' heads, which were never raised out of the water.

Quoy and Gaimard named their discovery *Delphinus rhinoc-*

121

eros. Their countrymen Michel Raynal and Jean-Pierre Sylvestre, who recently revived the mystery of this "rhinoceros dolphin" with an article in the journal *Aquatic Mammals,* note that, without a specimen, we can't be certain this strange cetacean should even be classified as a dolphin.

Any such uncertainty is not a reason to discount the sighting. Cetaceans' dorsal fins are made up of tissue only, without the supporting bone present in the fins of fish, so there is no practical reason why some species couldn't have two of them. Indeed, a freak example of the common dolphin, sporting an extra dorsal fin well back towards the tail, was observed in a school of normal specimens off Cornwall, England, in 1857.

As mentioned, this is not the only sighting. The Italian naturalist Antonio Montigore had reported the same or a similar dolphin al-

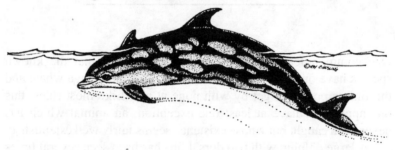

most a century earlier from the other side of the world.

If the species seen by Quoy and Gaimard still exists, then someone, someday, will bring one home for proper examination, and a striking new dolphin will appear in our textbooks. Perhaps, for better or worse, it will eventually appear at Sea World as well.

The Marozi

Africa boasts the world's most spectacular population of feline predators. Intriguingly, not all of these may have been catalogued.

There are a number of alleged African mystery cats, but the best reports come from the mountainous areas of Kenya. The beast involved appears to be a kind of small lion covered with prominent gray-brown spots. The male's mane is sparse and insignificant. To the local Kikuyu tribe, the spotted lion is the *marozi,* an animal distinct from the ordinary lion, *simba.*

The strong local tradition concerning the marozi is backed up by reports of European officials like game warden R. E. Dent, who saw four small spotted lions cross the path ahead of him in 1931. A short time later, a farmer named Michael Trent shot two such animals, a male and a female.

Naturalist and writer Kenneth Gandar Dower was sufficiently intrigued to go hunting for the animal. He found very clear pawprints, but was never able to track the elusive cat down. However, he returned to London with the skin of Trent's male lion and one skull. This evidence was examined by a prominent mammologist, Professor Reginald I. Pocock of the British Natural History Museum.

Pocock reported that the skin was smaller than normal for East African lions, about the size of a normal adolescent lion a year or so short of being full-grown. He described the mane as "small," but attached no significance to this, noting that some ordinary lions are almost maneless. The skull definitely belonged to a juvenile, either a female or an unusually small male.

It was the large, jaguar-like spots on the hide which made this "a remarkable specimen." Most lions have faint spots when they're born, but these fade quickly and never form the distinct pattern shown by Trent's skins. Pocock agreed with Gandar Dower's assessment that a unique race of lion was involved, although he didn't venture an opinion on exactly how to classify the beast.

The professor also wrote that he had previously examined two skulls, which he described as a male's and a female's, apparently belonging to marozi. These were from adult animals but were smaller than the smallest "ordinary" lion skulls.

Another hunter, Major W. R. Foran, later wrote that he had killed two spotted lions in 1906, but had assumed they were only freak individuals and had not saved the evidence.

One zoologist, Dr. Bernard Heuvelmans, notes that all marozi reports and remains come from mountain forests, which are not a

normal habitat for lions. However, smaller size and a dappled coat would be logical adaptations to such an environment, and he believes a distinct subspecies of lion has evolved.

The marozi is certainly rare. Indeed, no good reports have been made from Kenya since the 1950s, and the whole business might be dismissed as folklore if not for the hard evidence.

Writing in 1937, Gandar Dower suggested the marozi might not exist for long: that the plains lions, driven into the mountain forests by human settlement, might interbreed with their spotted cousins until the latter disappeared. Perhaps that's what is happening, although no intermediately spotted lions have been reported. Or it could be the marozi is still there, keeping to the most remote and inaccessible areas.

The skin of the female specimen, along with a single skull, remains housed in The Natural History Museum in Britain. The curator, Daphne Hills, reports that the skull on file belongs to a young animal of unknown sex. This illustrates the problem with old evidence: we can't be certain this is even one of the skulls Pocock described, and no one knows where the other skin is.

Concerning the skin, Ms. Hills believes too much has been made of it. She agrees the spots were retained later than usual, but suggests that this merely represents an aberration. Other zoologists, such as Dr. Karl Shuker, continue to disagree, with Shuker pointing to the strong local tradition and other sightings of adult spotted lions. Both Hills and Shuker note that reports of small spotted lions come from parts of Africa besides the Kenyan mountains, but they disagree on whether this is evidence for or against a specialized race of lions being involved.

The remains still housed in the Museum vouch for this handsome cat's one-time existence but do not settle the question of its identity. The marozi certainly does seem a worthwhile subject for a future expedition.

MacFarlane's Bear

In 1864, two Inuit hunters in Canada's Northwest Territories killed an "enormous" yellow bear. Naturalist Robert MacFarlane obtained the bear's skin and skull, and shipped the remains to the Smithsonian Institution, but no one at the time realized how unusual this specimen was.

Decades later, Dr. C. Hart Merriam found the evidence in storage in "the nation's attic" and took a closer look at it. He realized that MacFarlane's specimen, a nearly-grown female, was something unique. Of the known species, it most resembled the grizzly bear, but Merriam was convinced this was no grizzly. Beyond the peculiar coat color, there were differences in the skull and teeth separating the animal from the grizzly (or brown) as well as all other living bears. Merriam thought the teeth showed characteristics similar to some prehistoric bruins, and he described the specimen in 1918 as a new species and genus, *Vetularctos inopinatus*, calling it the "patriarchal bear."

While there were Inuit stories about such strange-looking bears, no other hard evidence has turned up in the last 130 years. Suggestions concerning the origin of MacFarlane's bear include a freak grizzly, a grizzly-polar bear cross (something known from zoos, although not confirmed in the wild), or a surviving representative— maybe the very last—of an ancient breed.

A modern polar bear specialist, Dr. James Halfpenny, doubts the idea of a "throwback" grizzly but is open to the hybrid theory. Unfortunately, no one to date has properly compared the skull to that of a known hybrid. In the vaults of the Smithsonian, the evidence of this mystery still awaits a final solution.

Giant Eels?

The famed "sea serpent," if it exists, is definitely *not* a serpent. However, many reports, such as the famous encounter by H.M.S. *Daedalus* in 1846 and a collision reported by the steamship *Santa Clara* in 1947, could be explained if we had hard evidence that giant eels were lurking in the oceans.

Perhaps we do.

Eels and some other fish pass through a larval stage where they are known as leptocephali. A normal eel leptocephalus measures about four inches in length. The largest known eels, such as the moray and conger, may grow as much as eleven feet long from this diminutive beginning.

In 1930, however, pioneer oceanographer William Beebe spotted a ten-inch leptocephalus from his bathysphere. That was startling enough, but that same year a Danish research ship collected a leptocephalus measuring six *feet*. A conger eel larva has 150 rudimentary vertebrae: the giant had 405. Another leptocephalus, three feet long, was caught off New Zealand in 1959.

The Danish scientists believed they had the larva of a giant eel and speculated the adult could be ninety feet long. Those observing such giant eels may have failed to identify them due to certain peculiarities of eel behavior.

Sea serpent reports often describe animals that swim by vertical undulation (impossible for a fish or snake) or which display a long neck. The conger eel has been known to swim on its side at the surface, creating the impression of a creature undulating vertically. On other occasions, this eel swims at high speed with the head and forebody held out of the water, making it look like something with a small head on a long neck.

This latter trait suggests a possible explanation for the most authoritative "sea monster" report on record. In 1905, observers on the yacht *Valhalla,* cruising off the coast of Brazil, spotted a dorsal fin about six feet long and two feet high. Then a small head on a neck seven or eight feet long rose in front of the fin. The creature's color was mainly dark brown, turning to white on the underside of the neck, and a "good-sized" body could be seen under the water.

This incident was unique because the observers were two experienced British naturalists, Michael J. Nicoll and E. G. B. Meade-Waldo, Fellows of the Zoological Society. They published an account of the sightings in that learned body's *Proceedings.*

While Nicoll guessed the creature involved was a mammal, this was admittedly only a conjecture. He did not report seeing definite evidence, such as fur or mammae. The sighting fits a giant conger-like eel well enough, especially since the central part of a conger's body thickens considerably, making the slim forebody look even more necklike. The only non-monstrous theory has come from Richard Ellis, a leading writer on marine life, who suggested it was a squid swimming tentacles-first with one arm held above the surface. This may not be impossible, but it's certainly hard to picture a squid behaving this way, and harder still to come up with a reason why it would even if it could.

There are reports which specifically describe giant eels, such as the twenty-foot creature seen off Britain by the crew of the German ship *Kaiserin Augusta Victoria* in 1912, but the 1905 case remains the only closeup sighting by scientists.

To confuse things further, the giant larvae may not be eels at all.

Two additional leptocephali, four and eleven inches long respectively and very similar to the New Zealand specimen, were identified in 1970 as members of a different group of fish, the noticanthiforms. These include the "spiny eels," eel-shaped bottom dwellers previously thought to measure about eight inches. Unlike an adult eel, which grows explosively from its larval size, known species of noticanthiforms stay about the same length.

Only the collection of an adult specimen will resolve the ambiguity. Are all of the leptocephali only evidence of rather ordinary fish, or could the Danish specimen, at least, still prove to be a clue to something fantastic? Danish oceanographer Anton Bruun, who netted the first leptocephalus, put it this way: "I am a man who rather believes in sea serpents."

The *Albatross'* Catch

That there are fish no scientist has yet described is a fact no one doubts. To illustrate the point, consider two reports from the Pacific Ocean, related by scientists almost eighty years apart.

In 1888, the American research vessel *Albatross* was trawling for specimens off the coast of Chile. One fish that tumbled out of the net caught everyone's attention. It was a previously unknown, primitive-looking creature, about five feet long, with two dorsal fins. (This brief description may remind you of a coelacanth, but the *Albatross'* fish was very different, having a more streamlined body and lacking the coelacanth's trademark lobed fins.)

The scientists on the *Albatross* photographed the fish, then apparently left it unattended on the deck. A neatness-minded sailor threw the fish overboard. It's not recorded whether the scientists threw the sailor overboard, but no one has seen the strange fish since. Only the photograph remains.

Time and technology change, but the mystery of the sea remains. Two researchers who manned the submersible *Deepstar 4000* on a

1966 probe of the eastern Pacific, for example, came face to face with a strange and awesome denizen of the deep. They were cruising at a depth of 4,000 feet in the San Diego Trough when a dark-colored, mottled fish they estimated was thirty to forty feet long swam right up to the eighteen-foot sub. The fish surveyed the craft and its startled inhabitants with eyes "as big as dinner plates," then moved off, apparently having decided the *Deepstar* wasn't edible.

Photographs taken later in the same area by a robot camera showed a large fish identified as a rare Pacific sleeper shark. If that was what the two oceanauts saw, it would be, by far, the largest sleeper shark ever reported.

The witnesses, pilot Joe Thompson and oceanographer Dr. Eugene LaFond, don't accept that their visitor was a shark. Consid-

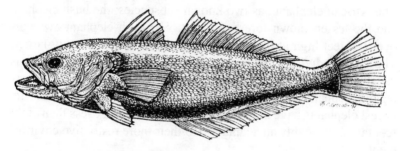

ering that both men described a round tail like a grouper's rather than a sharklike tail, it's quite possible they saw something even more interesting: an entirely new and gigantic species.

That doesn't exhaust the reports of strange fish prowling the abyss. For instance, a peculiar shark, six feet long with bulging eyes, was reported from a French bathyscaphe off West Africa at 13,000 feet.

We have certainly not found all the sharks in the world. A reference published in 1986 listed fourteen species known from a single specimen each. This is significant if one remembers that the difference between one specimen and none is usually mere chance. That was the case, for example, with the Antarctic sleeper shark. This species was described in 1939 from a lone carcass, eight feet long, that washed up on frigid Macquarie Island in 1912.

What else may be encountered in the depths? *Alvin* pilot Mac

McCamis had a glimpse of something he didn't want to see any closer. He called it "this monster...kind of shook me up. This was a living creature...I seen at least forty or fifty foot of it."

Someday, when science develops the technology to thoroughly explore the deep oceans, we will identify these intriguing creatures...and doubtless many more.

Is There a Pygmy Elephant?

The African elephant has two known subspecies: the bush elephant and the lesser-known forest elephant. The bush elephant averages over ten feet high at the shoulder, but may be considerably larger. The 13'2" individual mounted in the rotunda of the Smithsonian is usually rated among the all-time champs, and a thinner, longer-legged elephant killed in 1978 stood nearly fourteen feet tall in life. Forest elephants have more rounded ears and average less than eight feet tall, presumably an adaptation to their more restrictive environment.

In 1906, Professor Theodore Noack formally described a third subspecies—the pygmy elephant. He based his argument on a tiny specimen then living in the Bronx Zoo. This animal apparently did not read Noack's description, and inconsiderately grew larger. When it died of disease in 1915, it was six feet, five inches tall, close enough to normal forest elephant size to throw Noack's identification into doubt.

The pygmy elephant, however, proved to be a question defying a quick resolution. Noack's sort-of-pygmy wasn't the only abnormally small elephant caught. Another specimen was obtained in 1911, two more in 1932, and a fifth in 1948. All these elephants were under six feet high. Researchers in the Congo discovered the Uele tribe (pygmies themselves) had a different name for the pygmy elephant than for the forest elephant, and the government of that country even put out an official game-law guide listing the "dwarf elephant" as a rare species and demanding a stiff license fee for hunters wishing to shoot one.

Conservative zoologists held, and most still do, that pygmy reports involve dwarf or juvenile forest elephants. The reported range of the pygmy is basically the same as for the forest elephant, although a preference for swampy areas is alleged.

Mistakes involving juveniles certainly happen, but captured and killed pygmy specimens usually have full adult tusks. Some, judging by the wear of their teeth, are senior citizens. A female examined by a French zoologist actually had a full-term fetus in its womb.

What's needed to establish the pygmy's uniqueness is evidence of entire herds of mini-elephants. A hunter who killed a specimen in 1957 reported it was one of a group of twenty-one similar animals, and three pygmies together were recently filmed in the Central African Republic.

One zoological writer, Richard Carrington, has suggested a sort of interim solution. In a 1959 book on elephants, he theorized there may simply be a geographic race of the forest elephant which runs smaller than normal.

David Western, former director of Wildlife Conservation International, thought he solved the pygmy elephant question for good in 1986. Western, accompanied by Dr. Richard Barnes and writer Peter Matthiessen, surveyed the forest elephant's habitat and discovered that some juvenile forest elephants prematurely develop adult-sized tusks. That expedition also found there was considerable interbreeding between the forest and bush elephants, creating a number of variations with mixed characteristics. These hybrids presumably explained the local inhabitants' belief that two types of elephant roamed the forests.

But only three years after Western closed the scientific lid on the pygmy elephant's coffin, two German zoologists pried it off again. In 1989, Martin Eisentraut and Wolfgang Bohme presented the case for the pygmy's identity, not just as a new subspecies, but as a full species. They reviewed the reports and film and added a biometric analysis of pygmy elephant skulls which, they argued, showed significant differences from the skulls of adult or juvenile forest elephants. Their attempt to revive the pygmy elephant hasn't convinced everyone, but hopefully it will spur further expeditions to settle the matter.

It's worth noting that there *were* pygmy elephants within historical times. One species, three to five feet tall and slightly built by elephant standards, died out in Sicily sometime before the birth of Christ. A leading Thai zoologist, the late Dr. Boonsong Lekagul, believed that a pygmy Asian elephant less than five feet tall lived in his country's southernmost region less than a hundred years ago. A surprising variety of other pygmies, some related to the modern elephant and some to the mammoth, existed in the more distant past, including species inhabiting Crete, Malta, Cyprus, the Philippines, Santa Rosa island off California and several islands in Indonesia. Some of these micro-pachyderms were only three feet tall.

A final note: there is a breeding project underway in Thailand to *create* a pygmy elephant for use on teak plantations. The goal is an easily handled (and tuskless) work animal. (If this seems absurd, remember that human breeders—or meddlers—started with the wolf and managed to wind up with the dachshund.)

None of these oddities bear a direct relationship to the alleged African pygmy, but they do show that elephants don't *have* to be truck-sized.

With all the effort expended by zoologists to study African animals, plus all the effort expended by hunters to kill them, the idea that we have overlooked an elephant seems incredible...but it just might be true.

The pygmy elephant has a stronger case for existence than another alleged miniature, the pygmy gorilla. As in the case of the pygmy elephant, we have specimens but no agreement on what they represent.

The average male gorilla stands about 5'9" in his usual knuckle-walking posture. Females are more than a foot shorter. However, a pygmy species, *Gorilla mayema,* was described from the Congo in 1877. Since then, two undersized zoo specimens captured in Gabon have also been cited as evidence. One of these was only two feet tall at the mature age of eleven.

Most primatologists believe these specimens were all abnormal individuals rather than proof of a unique species. No one has reported seeing an entire family of such apes. Unless someone does,

Gorilla mayema will languish in the zoological back room reserved for animals that probably were never real.

Big Fish, Little Fish

No one has discovered more of the sea's marvels than Jacques Cousteau and the crew of his famous research vessel, the *Calypso*. Among the countless unusual fish and other creatures these divers have encountered, two finds stand out as unique—indeed, almost unbelievable. One concerned freakish specimens of a known species, the other an almost comical little fish which remains unidentified.

The first discovery was the truckfish. Two of Cousteau's men, diving on a wreck off the coast of Africa, returned to the ship laughing and nearly incoherent. They eventually explained they were still trying to believe what they'd seen—a colossal dark green fish about thirteen feet long and eight high. Unable to identify it, they called it the "truckfish."

Leafing through reference books, the divers finally decided the truckfish was a wrasse. But they had never seen a wrasse more than three feet long, and the sea explorers were at a loss to suggest what made one into such a monster.

Cousteau didn't specify which of the 600-plus species in the wrasse family they thought the truckfish might belong to. The common types found along the African west coast, the ballan and cuckoo wrasses, normally measure less than two feet: the tautog of the Atlantic may reach three. That question aside, if we grant three feet as the length of a normal wrasse and assume a conservative twelve feet for the truckfish, the giant had a total mass of 4 x 4 x 4 = 64 times normal size.

Later dives discovered a smaller wrasse, eight or nine feet long. This second find was, inevitably, nicknamed the "pickup" truckfish, but there were still no clues to explain the wrasses' growth to science-fiction proportions. The truckfish remains a mystery of the sea.

The second discovery was made in 1955 at Assumption Reef in the Indian Ocean. Cousteau described this area as a paradise of crystal-clear water and incredibly rich marine life. One diver, Emile Robert, claimed to have seen a small fish whose body was covered with perfect red and white squares. A fish with such absurd coloration was too much for even the underwater pioneers of the *Calypso* to swallow, and the comments Robert heard from his fellow divers mainly involved suggestions of nitrogen narcosis.

Days later, while several divers were down preparing for a film sequence, Robert began whooping and pointing. Cousteau wrote, "There sat a fish three inches long with a body pattern of perfect squares like a tiled floor. After that, I think I would have believed a man who told me he had seen an octopus wearing a derby hat and smoking a cigar." Marine biologists Cousteau consulted had never heard of this "checkerboard fish."

Cousteau had one more sighting of his own that reminded him anything was possible. Three thousand feet down in a bathyscaphe, he spotted a fish "twenty inches long and shaped exactly like a draftsman's triangle. It was the shade and thinness of aluminum foil with a ridiculous little tail."

In the oceans, it seems, even the absurd is routine.

The Imitation Sea Monster

If there are large unknown animals in the ocean, why aren't their carcasses found washed ashore? One answer to this reasonable question is that scavengers often dispose of the bodies of anything that dies. After all, no megamouth sharks drifted ashore before that first fortuitous catch in 1976. Also, many animals sink when dead. This is true of most whales, for example: whalers had to inflate their kills with compressed air to keep them afloat.

Still, "monster" carcasses do turn up on the world's beaches. Most of these are eventually traced to one of nature's little jokes, the basking shark.

This harmless shark reaches at least forty feet in length, and the live animal doesn't much concern us here. When dead, however, the basking shark decomposes in a most peculiar fashion. The lower jaw and the gill section drop off, the lower lobe of the tail disappears, and what you have when such a partially decayed carcass reaches shore is something that looks very much like a creature with a small head on a long neck.

These imitation plesiosaurs have caused a great deal of consternation. They have also made good tourist attractions, as in the case of a Massachusetts carcass found in 1970 that was served up in a local restaurant as sea monster stew.

In 1977, a Japanese fishing boat dredged up a carcass strongly resembling a plesiosaur, and reports that the sea monster mystery had been solved flashed through the world's media. The subsequent news that it was just another shark didn't receive nearly as much ink.

The last word on this subject must go to a correspondent who wrote to the *Manchester Guardian Weekly* in 1942 concerning a "monster" carcass recently identified as a basking shark. The letter writer, known only as "Lucio," voiced obvious disappointment:

> Yet again the doubting Thomas
> takes our precious monster from us
> and proceeds once more to bomb us
> With disclosures stern and stark,
> Lo! our portent meteoric
> doped with dismal paregoric
> sinks from monster prehistoric
> To a common Basking Shark.
>
> When we thought we had before us
> an undoubted something-saurus
> from the days when all was porous
> In the world's well-watered dish,
> These confounded men of science
> setting fancy at defiance
> go and put their cold reliance
> On an unembellished fish.

We need something more terrific
than these learned lads specific
I defy their scientific
And uncompromising quiz,
Their pretensions need unmasking
here's a question for the asking—
how could any shark go basking
With the weather what it is?

The Kamchatka Giant Bear

There are two known species of giant bears in the world. One is the polar bear, an outsized specimen of which stood over eleven feet tall and weighed a reported 2200 pounds. The other is the brown bear, varieties of which are known as the grizzly, the Kodiak, the Peninsula, and the Kamchatka bear. As we have seen, the question of a third large bear, MacFarlane's, is still shrouded in uncertainty.

Could there be yet a fourth? In 1920, Swedish zoologist Sten Bergman was shown the skin of what he suggested was a giant, black-furred variety of the Kamchatka bear. Dr. Bergman, who spent two years in Kamchatka studying the local wildlife, wrote that the specimen "far surpassed" any bearskin he had ever seen. Interestingly, the black bear's pelt was short-haired, unlike the long, shaggy coat of the normal Kamchatka bear. Bergman's report on the bear also included a description of an outsized pawprint, fourteen and a half inches by ten, and measurements of an equally outsized skull.

No specimens of this bear have been collected since Bergman wrote in 1936, so it is normally considered extinct. Still, in 1960, survey workers in Alberta, Canada found a valley inhabited by a huge strain of grizzly bears no one knew about. The bears in this isolated valley averaged 1,000 pounds in weight, compared to 600 for the normal adult grizzly.

With this example in mind, should we dismiss the prospect of rediscovering Bergman's bear? Terry Domico, in his 1988 book *Bears of the World,* notes that much of the Kamchatka Peninsula has long been closed off for military reasons. He also reports that a former Soviet official who did have access recently told him the giant bears were still reported.

Domico also suggests the giants are a variant of the brown bears living on the Siberian mainland, but there is no way to arrive at a definitive classification without a specimen.

Does the black giant still survive? Or was Bergman's bear the last of an unsuccessful race? We know this awesome creature existed barely seventy years ago. The largest land carnivore on earth may still be awaiting its rediscovery.

Freaks, Hybrids, and Fakes

Everyone agrees that many reports of unknown animals stem from an observer's honest mistake. Sometimes this just means the witness didn't get a good enough look at the animal involved. In other cases, a report may be based on a strange mutation or variation of a known animal. Accordingly, it's worthwhile to take a look at just what kind of oddities nature—or deliberate human action—may produce.

True albinos—entirely white animals completely lacking in skin pigmentation—are known in species ranging from humans to alligators. The opposite of this condition is melanism, which produces a solid black animal.

Albinism or melanism may be incomplete, affecting only part of the animal and shading into its natural color. Other less-drastic genetic variations include blue-eyed "white tigers," in which the body fur is white but the normal pattern of black stripes is present. These stunningly handsome cats are so popular that some zoos breed them deliberately. Fujian province in China is the reported home of occasional deep blue tigers, which must be strange-looking animals indeed. Zebras may produce "negative" examples, with black bodies striped thinly with white as a result of the normal color pattern being reversed. There is even a case of a *spotted* zebra, black speckled with white markings.

Turning from coloration to other characteristics, there is one case of a goat, born with hind legs only, which adapted and learned to move with a kind of hopping motion. Zoologist Maurice Burton wondered whether such a creature might have given birth to the legends of Pan and other goat-men. The opposite peculiarity was displayed by a humpback whale caught off British Columbia in 1919. This animal had vestigial hind legs four feet long! A two-headed shark was once caught in Australia's Botany Bay: one can only imagine the reaction of the fisherman who hauled it in.

Zoologists have recorded three-antlered deer and three-horned rhinoceros. One rhino at the San Francisco Zoo actually had a small horn growing from its *shoulder.* Hornless rhinos have also been collected.

What of the legendary unicorn? Unicorn goats have been produced deliberately by transplanting the horn buds when the animal was young: the Ringling Brothers circus has drawn crowds (and criticism) by exhibiting such goats.

An American biologist, Dr. W. Franklin Dove, performed such an operation on a male calf in 1933. The bull grew one large, straight horn, which it used to good advantage in sparring with other bulls. It's possible that some ancient cultures learned similar tricks, and may have produced unicorns on demand. As a final and very peculiar note on the subject of horns, jackal skulls sporting a small horn at the front or back of the head have been collected from Sri Lanka.

A strange animal might also be a hybrid. Species by definition don't normally interbreed, especially in the wild. But exceptions do occur, especially when an animal's population is dwindling and an individual is hard pressed to find a mate of its own kind.

The famous (or infamous) northern spotted owl has taken to breeding with the more common barred owl. The result, a dappled brown-and-white bird, has been nicknamed the "sparred owl." The hybrid is considered a nuisance by environmentalists making the case for the spotted owl's uniqueness and value.

Almost all dogs, wolves, and coyotes will crossbreed. Members of the genus *Equus* are similarly compatible, and so we have zebroids (half-zebra, half-horse) and zedonks (zebra-donkey crosses) as well as the familiar mules. Captive big cats have produced lion-tiger hybrids, lion-leopard crosses called "leopons," and jaguar-leopard crosses, one of which was in turn bred to a lion to produce what looked like a slender lioness with brown spots. There is a single case of a tiger-leopard hybrid killed in the wild. The world's only "litigon," three-fourths lion and one-fourth tiger, died in an Indian zoo in 1991.

Three odd-looking dolphins which washed up on an Irish beach in 1940 are believed to be hybrids between the bottle-nosed dolphin and Risso's dolphin, two species which weren't considered closely related. In a Japanese aquarium, a bottlenosed dolphin and a false killer whale produced a calf: the parents are both in the family Delphinidae, but not in the same genus, and the bouncing baby was quite a surprise.

Among the primates, siamangs and gibbons have produced viable offspring called "siabons." Large, odd-looking chimpanzees, called kookalamburras, have been blamed on chimp-gorilla matings, although most authorities consider them merely a robust race of chimps. Rumors of human-ape crosses have been around forever, but are apparently only stories.

It should be remembered that the less closely related animals are, the more likely an interspecies mating is to fail or to produce sterile offspring. One factor in this is the number of chromosomes: if the two animals have an equal or close number, the chances for successful hybridization are increased. For instance, the domestic horse has sixty-four chromosomes. Przewalski's horse, discussed in Section II, has sixty-six. The two interbreed without difficulty, producing hybrids with sixty-five chromosomes which can mate successfully with either of the parent types.

Before leaving this subject, we should mention the deliberately faked animals. Rays have for centuries been mutilated into strange-looking creatures touted as "sea-devils": the fakes themselves are known as Jenny Hanivers, a term whose origin is a mystery. Fake photographs have been offered as evidence for the tatzelwurm, an unproven but not improbable reptile or amphibian from the Alps, and for an eel-like sea monster from Australian waters.

The best-known apparent hoax came in 1920, when a geologist

named de Loys emerged from the Venezuelan jungle with a clear photograph of the body of what he claimed was a newly discovered ape. A few scientists accept the photo, but to most the animal looks like a spider monkey, with the tail cut off and a few other alterations made.

Interestingly, comparison with the crate the body is propped on in the picture indicates a very large and robust spider monkey, perhaps five feet tall. That implied a size record or even a new subspecies: this may be a case where a hoax obscured a find of genuine scientific interest.

These are some of the possibilities we need to keep in mind when evaluating claims based on only a few specimens. Mother Nature does not always play straight with science: sometimes she throws us a curve.

The Giant Octopus

An octopus with a thirty-two-foot tentacle span, the largest on record, is big enough to scare any scuba diver into switching his or her hobby to mountain climbing. The "official" records, though, may well be incomplete. An octopus straight out of a Jules Verne novel once washed up on a Florida beach, and there are photographs and tissue samples to document it.

The story began in 1896, when Dr. DeWitt Webb, a medical doctor and amateur naturalist, became fascinated with a shapeless mass, twenty-one feet long, that drifted onto the beach near St. Augustine. Webb believed the carcass was of a giant octopus, and he sent samples and photographs to Dr. A. Hyatt Verill, the leading cephalopod expert of the day.

Verill at first agreed with the octopus idea, then suggested the thing was instead a giant squid. He later changed his opinion to "the head of a sperm whale," despite the lack of bones. By the time the

140

eminent scientist made up his mind, the carcass had washed back out to sea and everybody forgot about it.

More than sixty years passed before two enterprising scientists, Forrest G. Wood and Joseph Gennaro, dug samples of the creature out of the vaults of the Smithsonian Institution. Their microscopic tissue analysis and a 1986 biochemical study by Dr, Roy Mackal came to the same conclusion. The lump on the beach *was* an octopus...and the tentacle span may have been 200 feet.

Those who agree include fishermen in the Bahamas, who accuse such "giant scuttles" of robbing their lines and traps, and Cuban fishermen, who have claimed encounters with octopi fifty feet across. More authoritative is the report of the U.S.S. *Chicopee*. In the spring of 1941, two destroyers dropped depth charges off Fort Lauderdale in search of German submarines. Sailing close behind came the *Chicopee*, whose crew investigated what looked like debris and discovered a titanic brown octopus, floating in an apparently stunned condition and no doubt wondering what had just happened.

In the 1990s, the controversy erupted again. Richard Ellis, in his 1994 book *Monsters of the Sea*, published a previously unknown photograph showing what looked like a tentacle on the St. Augustine lump. A team of four dissenting scientists shot back in 1995 with new chemical and microscopic analyses indicating the thing was a lump of skin and blubber torn from a whale. There was no identifiable DNA in the century-old samples, which was disappointing to everyone who hoped for a definitive resolution. Ellis and other octopus supporters were unconvinced. What part of a whale, they asked, could be torn off to produce a mass weighing an estimated five tons but not containing any bones, organs, fins, or anything else recognizable? Was it, as skeptics suggested, the entire skin? If so, how did it become detached from the whale?

Octopus giganteus may not be in the textbooks yet, but it hasn't exactly disappeared, either.

Kon-Tiki's Companions

It's hard to imagine a better platform for observing marine life than a slow, silent raft. The crew of Thor Heyerdahl's balsa *Kon-Tiki* discovered this was true when, shortly after World War II, they set out to test Heyerdahl's theory that South American natives could have spread their culture across the Pacific through raft voyages.

The *Kon-Tiki* became something of a floating island, festooned with seaweed and investigated by all manner of marine fauna. The six men of the raft's international crew observed countless sea creatures, including a titanic whale shark. Heyerdahl also recorded a blue shark he estimated was nearly twenty feet long, which would be a record for that sleek predator.

The voyage did not result in the formal description of any new species, but the crew did collect the first whole specimen of the weird deepwater predator called the snake mackerel. "Collect" may be too strong a word, since the *Kon-Tiki*'s two specimens of the fish obligingly jumped onto the raft at night.

The snake mackerel was previously known only from a few skeletal remains washed ashore in South America. The new specimens showed the eel-shaped fish was about three feet long in life. It was colored violet and blue and featured large black eyes and blade-like teeth. One drowsy crewmember studied this finned nightmare for a moment and shook his head. "No, fish like that don't exist," he decided.

Of special interest in this section, however, are the visitors that couldn't be identified at all.

Heyerdahl recorded that, on several nights, the raft was surrounded by "round heads two or three feet in diameter, lying motionless and staring at us with great glowing eyes." At other times, the crew spotted "balls of light" over three feet across, flashing on and off under the waves.

One night, a huge, phosphorescent form maneuvered back and forth under the *Kon-Tiki*. It appeared to change shape, then split into two and then three shining things, whose visible parts alone were estimated at thirty feet long. No features were visible, just the huge, vaguely oval backs of the three unknown animals, circling under the raft for hours without surfacing. Fascinated crewmembers hung lights over the side to lure the mysterious visitors up, but without results.

The crew also observed fish they couldn't name. One was described rather puzzlingly in the log as a "thick dark colored fish with a broad white body, thin tail, and spikes." Another was six feet long with a "thin snout, large dorsal fin near head and a smaller one in the middle of the back, heavy sickle-shaped tailfin." It swam by "wriggling its body like an eel." On one occasion, thirty of these were observed in a school.

Several times the raft passed "a huge dark mass, the size of the floor of a room," which stayed motionless as they drifted by.

Heyerdahl thought this last creature was a giant manta ray. Some of the other fellow travelers might have been enormous squid. It would be interesting to repeat this voyage, this time carrying modern sonar, night-vision equipment, and collecting apparatus. As it is, some of *Kon-Tiki*'s strange companions could remain mysteries for years to come, reminders that beneath the waves is something very much like another planet, where we are only visitors.

The King Cheetah

The cheetah is the fastest and most specialized of the world's big cats. It is also the most monmorphic: in other words, cheetahs are much the same all across their wide African-Asian range.

That makes the story of a strange cheetah-like cat even more intriguing. Early white settlers in Africa were told by local hunters about an animal called the *nsuifisi,* or hyena-leopard, marked with stripes down its back and large irregular blotches on the sides.

In the 1920s, naturalist A. L. Cooper obtained five skins of this cat. He sent one to a London zoologist, Professor R. I. Pocock, who pronounced it a new species—the king cheetah. Many zoologists disagreed, calling the cat just a local variety or mutation of the cheetah, and Pocock retracted his separate-species opinion in 1939.

Zoology generally forgot about the king cheetah for the next forty years. Then the first-ever king cheetah expedition, led by Paul and Lena Bottriel, penetrated the remote areas of Botswana to learn more about the peculiar cats.

The Bottriels were intrigued by the theory that the king cheetah might represent a newly developing variant. There are no records of king cheetahs before the nineteenth century. Equally interesting, there are no intermediate forms. A cheetah either has normal markings or the strikingly different king cheetah coat, never anything in between.

Examining every known king cheetah skin, the Bottriels found that king cheetahs were invariably large specimens and that the skin showed longer, silkier hairs than on a normal cheetah pelt. Microscopic analysis showed the structure of each hair to be very different from ordinary cheetah hair. In fact, it resembled leopard hair, although leopards will kill cheetahs and do not interbreed with them.

Of the approximately forty confirmed king cheetah specimens the Bottriels surveyed (counting live animals and skins), all came from the same region south of the Zambezi River.

To the Bottriels, all this points to a new geographic race of cheetahs emerging, one more appropriately marked for a forest environment than for the plains where the cheetah normally dwells. According to sightings, the king cheetahs are in fact more likely to inhabit forests. Since the plains habitat is shrinking, the king cheetahs may have an evolutionary advantage and be likely to increase even as the endangered cheetah species as a whole is declining.

Are we seeing evolution in action—the gradual emergence of a new cat? It's a fascinating question, but one that will take a long time to answer.

Some information has come in from South Africa's De Wildt Cheetah Research Center, where normally marked cheetahs gave birth to nine king cheetah cubs between 1981 and 1984. According to the Center's researchers, the king cheetah trait seemed to be an inherited mutation passed down mainly through the males. One of the males involved came from Namibia, far from the restricted area where all other kings have been bred. This shows that, while the trait may predominate in one area, it can crop up in other cheetah populations.

There is one more strange episode in cheetah history. In 1877, the London Zoo obtained a very odd-looking cheetah from South Africa. Rather than the usual black spots, this large, relatively short-legged cheetah had a pale coat marked with brown splotches. The fur was very dense and woolly, and the animal has become known as the "woolly cheetah."

At least one more woolly cheetah specimen was identified, coming from the same region. However, no such animal has been found since 1884. If the king cheetah is an evolutionary experiment in progress, perhaps the woolly cheetah was an idea Mother Nature explored without success and has now forgotten.

Legaut's Mystery Bird

In the western Indian Ocean lie the Mascarene islands. This chain, including the islands of Mauritius, Reunion, and Rodrigues, is—or was—the home to some of the strangest bird life in the world. The most famous example is the dodo, which looked more like one of the late Jim Henson's Muppets than it did a real bird. Human settlers and their animals drove the Mauritius dodo to extinction by 1700. Its relative on Reunion lasted several decades longer, and a few romantics suggest it lingers to this day.

Francois Legaut was the leader of a 1691 French expedition to colonize the small island of Rodrigues. He was also an amateur naturalist who described and sketched the native plants and animals. After two years on the island, Legaut returned to France, where in 1708 he published an account of his experience.

In his book, Legaut described a long-legged rail or crane that lived mainly on Mauritius, although he met one stray specimen on Rodrigues. It was entirely white except for red patches under the wings, and it stood six feet tall. Legaut provided the only known illustration of this bird, which he called the Gean.

Was the Gean a hoax? A remnant of a now-extinct species? A lost flamingo? All we know is that no one has found it since Legaut's report.

The problem bird looked something like a goose perched on long legs with outsized feet. Legaut described it as a marsh-dweller and said it was often killed by the dogs the settlers introduced. The lone specimen found on Rodrigues was killed by Legaut himself. Unfortunately, he ate his catch rather than preserving it.

No other early visitor left any report of this striking bird, which has been scientifically labeled *Legautia gigantea,* the giant water hen. Legaut's descriptions and illustrations of other species, however, mark him as a reliable observer. One bird he described was the

Rodrigues solitaire, a relative of the dodo. His account of such a bird was doubted for a long time, but fossilized and unfossilized remains have since been found that establish the solitaire's existence.

There rests the matter of what must have been a very memorable creature. There are some little islets in that area which remain almost unknown: uninhabited, rarely visited, and not even precisely charted. Accordingly, the story of Legaut's strange bird, like that of the dodo itself, has a few blanks that may yet be filled in.

The Titanic Tunicate

This section might be subtitled "big squishy things." Jellyfish and other aquatic invertebrates may not be of interest to most people, but some of them appear to be a lot more interesting—and considerably larger—than we thought.

Some seagoing invertebrates, like the dangerous Portuguese man-of-war, are actually colonies, floating communities whose members have various functions. In another phylum are some lesser-known colonial creatures called the tunicates. The members of one order, the pyrosomatids, form aggregations shaped like a cylinder or cone. Seawater is ejected from the open end, giving these colonies an undulating mobility that might explain some sea serpent reports. Recent years have brought observations of colonies over thirty feet long. Another order, the salps, consists of simple individual animals, usually six inches long or less, that may join to form narrow chains stretching to eighty feet.

There are observations of even stranger invertebrates. The first was recorded over a century ago by the captain and the ship's surgeon of the British vessel *Nestor*. In the Malacca Straits, on September 11, 1876, the *Nestor* encountered an awesome creature so large it was initially mistaken for a shoal.

According to the official statements sworn to by Captain John Webster and Dr. James Anderson, it wasn't until the observers realized the massive object was moving slowly through the water that the idea of a living animal crossed their minds. But what sort of animal? Reaching for comparisons, the captain and the surgeon both suggested a gigantic tadpole or other amphibian, even though no eyes, mouth, or fins were observed. The monster was pale yellow, marked along its entire length with bands of black. The region they thought of as the "head" was about twenty feet long, followed by fifty feet of "body" and an estimated one hundred fifty feet of "tail." The ship cruised right next to the thing for half an hour before pulling away.

This description does not belong to a tadpole or any other kind of vertebrate. Fortunately, the account provides some important clues. The thing was completely featureless, did not react at all to the nearby ship, and was made of a substance the surgeon described as "gelatinous" and "flabby."

Accordingly, biologist Roy Mackal suggests the *Nestor* met the largest tunicate colony on record. Dr. Anderson's inability to identify such an incredible thing is understandable, to say the least.

A more recent report concerns a smaller but equally bizarre creature. The U.S. Fish and Wildlife research vessel *Challenger* was operating off New Jersey in 1963 when the scientists on board observed an almost transparent creature about forty feet long and only five inches wide. Dr. Lionel Walford, one of the observers, thought it was an eel-shaped jellyfish relative known as a Venus' girdle. However, the largest known Venus' girdle is approximately five feet long. This giant form has not been reported since.

Aquanauts probing the depths in submersibles have spotted their share of unidentified spineless objects as well. Victoria Kaharl, in her book chronicling the *Alvin*'s adventures, recorded a number of such encounters. Biologist Frederick Grassle was an *Alvin* passenger when he reported a potato-sized creature of unknown classification he called an "undulating hairbrush." Another scientist described a "blue lampshade," hollow inside and about three feet across. A ctenophore, or comb jelly, about five feet across was a startling observation. Then there was the small six-sided creature observed in 1986. At a loss to classify it, the witnesses christened it "the Chinese checkerboard animal."

Finally, a Shell Oil robot inspecting a drilling rig in the 1960s

148

encountered a most peculiar sight. A still unclassified type of colonial creature, some fifteen feet long, was recorded on film as it propelled itself through the depths by rotating its collective body, which was wound about with ridges like a huge flexible screw. Shell workers named the thing "Marvin."

Somewhere in the ocean vastness, there must be more forty-foot Venus' girdles and more tunicate colonies the size of submarines. Some day, a lucky marine biologist will corral and properly classify such a creature, and at least one kind of "sea serpent" will leave the realm of folklore and enter the textbooks of science.

The Curious Kellas Cat

Britain, for a well-populated country, has a surprising number of hard-to-account-for reports of big cats padding around. Some, like the "Surrey puma," have sparked massive hunts which usually find nothing. Experts write most such cases off to misidentifications of dogs, feral tomcats, etc.

At least a few reports are due to escaped or released exotic pets. Iverness farmer Ted Noble, tired of losing sheep to an unseen killer, build a box trap in 1980 and caught an aging female panther. "Felicity," as she was named, spent her last five years of existence in a zoo. More recently, following sheep killings in Exmoor, naturalist Trevor Beer has tracked and photographed what looks like a black leopard.

In one case, however, something really peculiar turned up. Since 1984, at least eight specimens of a unique cat have been obtained from northern Scotland. The Kellas cat, as it's now known, has bristly black hair and strikingly enlarged canine teeth. It's taller and longer than a domestic cat, but not much heavier, thanks to its slight build. The limbs, especially the hind ones, are long and powerful. In the cat family, such a build indicates an animal adapted to run down its prey, as cheetahs do, instead of using the more typical feline stalk-and-pounce approach.

149

The Kellas cat's skull and other identifying characteristics show some traits of the known Scottish wildcat and some of the domestic cat. Accordingly, the consensus of zoologists who have studied the subject is that it's a hybrid, but of a most unexpected type. Previous known hybrids resemble the stoutly built, tabby-striped wildcat. The Kellas cat's unique build, and the number of identical specimens obtained, raised some scientific eyebrows.

Repeated hybridization can bring out latent genes and produce unexpected results. The question concerning the Kellas cat is the future. If there are enough of this type produced, and they begin to mate with each other, will they eventually breed true to type, resulting in a truly distinct animal? There are no reports so far to establish that these hybrids are fertile, but the possibility is intriguing.

Indeed, zoologist Karl Shuker, who has done most of the studies of the Kellas cat, has come across something even stranger. Two specimens of *Felis chaus*, the Asian jungle cat, have been killed in Britain since 1988. This handsome, tufted-eared feline outweighs the average domestic cat by better than two to one, but is known to produce fertile hybrids with our familiar housepets. Sightings of unidentified cats in the areas where these jungle cat specimens were found suggest to Shuker that a new hybrid cat may be in the process of establishing a permanent presence.

Just to confuse things completely, another Asian import, the leopard cat, has also been found in the wild in Britain. This animal, too, can interbreed with the domestic cat—in fact, such breeding has been deliberately arranged to produce a trendy housecat with a spotted coat. To sum all this up, a most interesting, if uncontrolled, genetic experiment is underway in a most unlikely place. It will be interesting to see, decades from now, just what kind of cats are roaming the fields of Britain.

Mr. Benchley's Monsters

Peter Benchley has made a good living from two of the most fearsome creatures of the sea: the great white shark and the giant squid. One question his novels raise in everyone's mind is whether animals the size he describes in *Jaws* and *Beast* are really lurking in the depths.

Answer: maybe.

Are there monster sharks like Benchley's twenty-five-footer? Richard Ellis and John McCosker, in their 1991 book *Great White Shark*, went to exhaustive lengths to investigate the conflicing claims about the great white's size. Many books list a thirty-seven-foot shark caught in 1930 off New Brunswick as the all-time record, but Ellis and McCosker found no evidence for this fish's existence. An oft-quoted Australian record of thirty-six-and-a-half feet is apparently due to a typographical error: the true length was sixteen-and-a-half feet. However, the authors do accept a 1987 Australian catch whose preserved jaws provide evidence of a shark over twenty-two feet long. In the same year, a specimen of about the same size was taken and photographed near the Mediterranean island of Malta. One expert, Dr. John Randall, attributed bite marks on whale carcasses found off Australia to a great white at least twenty-five feet long.

In short, Benchley's fish would be a near-record, but it's not beyond possibility.

There are reports from witnesses, including author and champion deep-sea fisherman Zane Grey, of great whites ranging from thirty feet to over forty feet long. There is, however, no physical or film evidence to back them up. There is also no hard evidence for the ultimate shark, a ghostly white monstrosity allegedly encountered off Australia in 1918. The awed lobstermen who saw it described a creature at least a hundred feet long which gobbled a whole string of lobster traps like popcorn. Dr. D. G. Stead, who questioned the witnesses for his book *Sharks and Rays of Australian Seas*, called some of the estimates of length "absurd" (one man said 300 feet!) but still concluded the men had met "a vast shark."

Misidentifications of the harmless basking shark, which reaches at least forty feet, may be involved in some cases. (It's been suggested that the thirty-seven-foot great white reported from 1930 was one such case of mistaken identity.) The equally inoffensive whale shark is even bigger, but has a distinctive spotted back unlikely to be overlooked.

A more romantic but unproven explanation is the possible survival of the great white's big brother, the prehistoric *Carcharodon megalodon*. Reconstructions of this animal, based on fossil teeth ranging up to seven inches long, vary from forty to ninety feet in length. Most authorities consider the larger figure inaccurate, but here is a fish that could certainly account for the stories of monster great whites. (By way of comparison, a seven-inch tooth is more than twice as large as any recovered from a modern great white.)

The only hard evidence to support such speculation turned up in the 1870s, when the research vessel *Challenger* dredged up two five-inch shark teeth from the Pacific ocean. While these, like all *megalodon* teeth found so far, were fossilized, they were dated at 11,000 and 24,000 years old. Eleven thousand years is a very short span on the geological calendar. Maybe those hoping to catch an all-time record maneater are not entirely without hope. However, as Chief Brody pointed out, they'll need bigger boats.

The giant squid was a fixture in legend long before it was known to science. "Far, far beneath in the abysmal sea...the Kraken sleepeth," wrote Tennyson, echoing the age-old stories of titanic squid lurking in the depths. It wasn't until the late nineteenth century that this outlandish beast gained full scientific acceptance as a real animal.

While not the island-sized creature of myth, it was certainly a giant. It could also be deadly: in March, 1941, a survivor of the torpedoed troopship *Britannia* was dragged off his life raft and killed by a hungry squid.

Just how giant is the giant squid? Well, there is no question that squid over fifty feet long (measuring to the tips of the two long tentacles) exist. We have stranded specimens to prove it. The largest definitely known squid, fifty-five feet long and weighing almost two tons, came ashore at Thimble Tickle Bay in Newfoundland in 1878. The *longest* specimen generally accepted is a fifty-seven-footer with very long tentacles from New Zealand.

British biologist Michael Bright has collected several reports of larger squids. One was caught under a log boom being towed along the Canadian west coast in 1892. When the raft was finally beached, the mangled squid was allegedly found to have tentacles stretching over a hundred feet. In the same area of Port Simpson, a dead squid drifted ashore in 1922. This one also boasted hundred-foot tentacles. In 1968, a Soviet whale-spotting helicopter reported a truly monstrous squid in the Indian Ocean. Bright does not cite an estimate of length, but says the arms were over three feet in diameter.

A mutilated squid estimated at a hundred feet long washed ashore in South Africa in 1926. A Newfoundland report from 1882 concerns a beached eighty-eight-foot squid, and a better documented and measured squid stranded in 1934 was seventy-two feet long. (The popularity of Newfoundland as a squid burial ground may have to do with warm and cold water currents colliding off that coast and with the imperfect buoyancy mechanism of the giant squid. A squid caught in water too warm will be unable to submerge, and one caught in water too cold will freeze.)

Stories like that of sailor A.G. Starkey, who reported seeing an Indian Ocean squid longer than his 175-foot ship during World War II, fall into the same category as giant great whites: they are not physically impossible, but most scientists quite reasonably demand to see a carcass, or at least a piece of one.

Pieces of impressively large arms or tentacles were often reported by early whalers when dying sperm whales vomited their last meals. At least two American ship captains reported tentacles with suckers "as big as a plate." In one case, these were attached to a dismembered arm forty-five feet long. (The suckers on the Thimble Tickle squid were about three inches across.) Unfortunately, no such specimen has been preserved. Circular scars left on the skin of predatory whales provide another form of evidence. Scars nearly five inches across have been measured by scientific observers, and there are whalers' reports (again, without preserved specimens) of scars as large as *two feet* in diameter.

Part of the problem is that we know even less about the giant squid than we do about the great white shark. We have only guesses about how many species exist, how long they live, and what their growth patterns are like. Without such basic data, it's hard to pass judgement on the question of maximum size.

A final obstacle is that, if such monsters do exist, there is no practical method of fishing for them. It doesn't matter how big your boat is if your quarry can just pluck you off the deck.

The Double-Banded Argus

The double-banded argus is a true mystery bird. The evidence for its existence consists of a single feather, now kept in the British Museum. The feather apparently belongs to a pheasant from the genus *Argus,* but its pattern can't be matched to feathers from either of the two known species. The formal description of the new species, based solely on this feather, was written by T. W. Wood in 1871.

The location where the feather was collected, unfortunately, is not on record. (Known Argus pheasants live in Borneo, Sumatra, and on the Malay Peninsula.) The feather could be from a single aberration, but one never knows.

Bird puzzles don't come only from exotic locations overseas. Sutton's warbler is an American mystery, a bird no one can agree on. Discovered in 1939, the species (if that's what it is) lives in the Middle Atlantic states. Almost nothing is known of its habits or range. The bird is so rare that a nesting pair has never been reported.

The species' validity is the key point in dispute. Depending on who you ask, the bird is a rare species (*Dendroica potomac*, to be precise), a new species just emerging, a subspecies, or a cross, perhaps between the northern parula and the yellow-throated warbler. The song resembles that of the former bird, the plumage that of the latter.

A similar aura of uncertainty surrounds Townsend's finch, collected for the first and only time in the early 1800s in Chester County, Pennsylvania. John James Audubon wrote a description of this species, but no one has seen it since. Some authorities believe the specimen was a freak or hybrid, while others maintain it's too distinct to be anything other than a genuine species.

A bird in the hand, it seems, is not always worth one in the textbooks.

Tracks in the Abyss

Ever since humans have gazed on the surface of the sea, they have wondered what forms of life moved in its depths. In the last century, we've employed progressively more sophisticated sampling devices to probe the floor of the deep ocean. Still, we cannot by any standard be said to be "familiar" with the abyssal regions.

When scientists began to photograph the deep sea bed, they found it marked by tracks of all kinds. Occasionally, a lucky frame would show an unidentified animal itself, like the small sluglike creature caught on film three miles down in the Indian Ocean. But mostly, there were just the tracks: a bewildering variety of footprints, ridges, furrows, and mounds left by strange denizens of the deep.

Fresh tracks have been found identical to those left by trilobites in rocks hundreds of millions of years old. This is not proof of living trilobites, but it is undeniably intriguing.

A ridgelike trail photographed two miles down in the Pacific marks the passing of an animal perhaps eight to twelve inches wide (a monster by abyssal standards). Next to the trail lie some sort of pellets about the size of golf balls. Fiddler crabs roll sediment into similar balls, but if this is the work of such a crab, it is a larger, deeper-dwelling species than any we know of.

Also from the Pacific come photographs of so-called "tread trails," about eight inches wide, looking for all the world like the track of a knobby truck tire. These appear from nowhere and break off just as suddenly. Sometimes they just stop, as if the trail-maker swam away. Other trails end in deep holes.

There are narrow ridges with leaflike impressions on both sides, giving the impression of a vine lying limp on the ocean floor. Another kind of trail consists of oval impressions, compared in size to a cat's paw, meandering in a single line. Both these types may be the work of an unknown fish plowing up the sediment for food.

Excrement is a common feature of the sea floor, left in trails and coils by invertebrates that suck in the sediment and filter it for nourishment. Some of the animals leaving these droppings are also unidentified. In the cold and dark of the great oceans, there are still secrets to be learned.

The Giant Gecko

New animals don't always turn up in their native habitats. The largest species of gecko that ever lived—a heavily built lizard about two feet long—was found in 1986, when someone uncovered a preserved specimen in the basement of the Marseille Natural History Museum in France. It is still the only known example of its kind.

The museum archives held only fragmentary records concerning this fascinating specimen. Investigating zoologists finally decided the gecko was probably shipped to the museum from New Zealand sometime in the nineteenth century. Inquiries revealed that New Zealand zoologists were unaware of the species. The native Maoris apparently *did* know about the animal, as their folklore included a forest-dwelling reptile called the kawekaweau.

Some partial skeletal remains have since been found in the gecko's homeland, but they don't answer the crucial questions. Is this impressive lizard living or extinct? If it is extinct, when did it

vanish? And just when and where was the French specimen collected, anyway, and by whom?

There are several recent sightings of a lizardlike animal in the northernmost forests of New Zealand's North Island. Two National Museum herpetologists recently interviewed the witnesses and surveyed the reported habitat. They classified some of the sightings as unexplained, but found no hard evidence the gecko was still alive. If the animal is extinct, the blame probably falls on cats, rats, and other alien predators introduced by early white settlers.

The question is not yet settled. North Island contains virgin rain forest whose canopy might conceal a viable population of the little-known reptiles. The giant gecko could have taken to the trees to avoid the new predators and may yet be looking down curiously at the humans trying to establish its fate.

Mysterious Whales

In an earlier section, we met the world's least-known whales. Could there also be whales we haven't identified at all? The answer appears to be "Yes."

What looked like a sixty-foot sperm whale with a high dorsal fin was reported by naturalist Robert Sibbald. (Sibbald was the first great authority on whales, and the blue whale is still known as Sibbald's rorqual.) The known sperm whale has no dorsal fin, just a sort of bumpy ridge. Sibbald described his species in 1692 as *Physeter tursio.* Other seventeenth-century mariners also reported this whale, mainly off the Shetland Islands.

Speaking of dorsal fins, we've mentioned that all known cetaceans have one at most. In September 1867, however, Italian naturalist Enrico Giglioli reported watching a large baleen whale with two dorsal fins off the coast of Chile. He observed it well enough to write down a detailed description and propose a scientific name, *Amphiptera pacifica.* Giglioli's whale was about sixty feet long and "gray greenish" with a grayish white underside. The dorsal fins were triangular and over six feet apart. Sightings of what may be other specimens of this whale were reported by fishermen off Scotland in 1898 and in the Mediterranean in 1983.

Philip Gosse, a prominent naturalist and nature writer of the nineteenth century, once watched a school of thirty-foot beaked whales in the North Atlantic for twelve hours. These whales, black on top and white below, with all-white pectoral fins, have never been identified.

Another mysterious beaked whale, a denizen of the eastern Pacific, has been spotted several times and was recently photographed off the coast of Mexico. About sixteen feet long, the animal is distinguished by a somewhat flattened head and a low, wide-based dorsal fin. Larger whales, presumably the males, are black with light-colored "racing stripes" down their sides: smaller individuals are uniformly bronze or grayish brown. This cetacean may represent the Indopacific beaked whale, previously known only from skulls, or it may be something entirely new.

Several Antarctic explorers, including the famous Sir James Ross, have described a twenty- to thirty-foot whale with a strikingly tall, slender dorsal fin. This animal was sketched in 1902 by explorer Edward A. Wilson, who reported observing a group of four such whales. All the observers were certain they were not seeing orcas (killer whales), which have a tall dorsal fin but whose striking color pattern makes them instantly identifiable. The unknown whale was seen again in 1911, and a strange whale possibly belonging to the same species was photographed by cetologists off Chile in 1964.

In 1981 and 1983, Russian mammologists described two new species of killer whales, or orcas, from the Antarctic. These "new" whales, *Orcinus nanus* and *Orcinus glacialis,* may represent separate reports about the same animals, which show yellowish rather than white undersides and are reportedly smaller than the standard *Orcinus orca*. Some experts attribute this color variation to a film of tiny creatures called diatoms rather than to the whales themselves, but there's so much we don't know about orcas that the subject remains open for discussion.

Sea captain and naturalist Willem F. J. Morzer Bruyuns, author of the *Field Guide of Whales and Dolphins,* reports he has observed an all-brown killer whale speckled with star-shaped white scars. All known killer whales, excepting the occasional albino, sport the same distinctive black-and-white color pattern. Bruyuns' twenty-foot "Alula whale" also presented a less streamlined appearance than the standard killer, with a higher, more rounded forehead. As many as four of these whales at a time passed Bruyuns' ship north of the village of Alula in Africa's Gulf of Aden.

In his forty years at sea, Morzer Bruyuns kept notes and made paintings of every cetacean he encountered. He reported several problematical observations besides the Alula whale.

For instance, the striped dolphin of the Mediterranean is named for a distinct pattern of black stripes that look like the bridle of a horse. Morzer Bruyuns described what looked like a shorter, stouter version of this animal, with no trace of the definitive black markings. Off Senegal, the captain reported large numbers of a brown and white dolphin about six feet long. To him it resembled the known bridled or Atlantic spotted dolphin, but again a characteristic marking, in this case a pattern of light spots, was absent. The white underside was also more distinct. Do these observations represent new races or subspecies of known dolphins, or something even more interesting?

Still more puzzling is what Morzer Bruyuns calls the "Illigan dolphin." This is a strikingly colored dolphin, brown with a pink underside and yellow flanks, seen off the Philippines in schools of up to thirty animals. In size and shape it resembled the solid black dolphin known as the melon-headed whale, but the color difference is so stark that the observer declined to classify it as merely a variation of that species.

Finally, Morzer Bruyuns in 1960 encountered a seven-foot dolphin that made no immediate sense to him. It looked like the Indian Ocean version of the bottlenosed dolphin, but with a longer, thinner snout. The basic silver-gray color was broken by dark patches in front of the dorsal fin and lighter gray skin aft of it. Bruyuns finally decided the lone animal might be a hybrid of the bottlenose and the Malabar dolphin, if such a mix "is at all possible."

Cetologists surveying the marine mammals off Chile in 1964 reported another puzzle: schools of "small, stout porpoises" about four feet long, brown on top and white beneath. Sea life writer Richard Ellis has noted that this sounds like the well-known harbor porpoise, but that species is thought to exist only in the Northern Hemisphere, so a new type may be involved.

It remains to be seen how many of these observations actually represent new species, but certainly some of them do. For cetologists, new discoveries are there for the making.

The Wayward Salamander

Amphibians, generally speaking, are animals of modest size. The largest living amphibian is the Asiatic giant salamander. Found in China, with a slightly smaller relative in Japan, it may approach six feet long. The largest known in North America is the hellbender of the Ozarks and Appalachians, which grows to over two feet.

In 1951, however, a most interesting article by Stanford University herpetologist George Myers appeared in the scientific journal *Copeia.* Myers reported he had examined a giant salamander caught in a catfish net in California's Sacramento River. The amphibian was definitely a *Megalobatrachus,* a member of the genus which includes the Asian giant salamanders. However, its color, a uniform dark brown with yellow spots, was "quite at variance" with the standard gray of Asian types, and "suggested the possibility of a unique California variation." The specimen was of modest size for *Megalobatrachus,* measuring only thirty inches long.

Unfortunately, the fisherman only allowed the specimen to be examined, not kept for further study. What became of the salamander, then living in a bathtub in its owner's apartment, is unknown.

Giant salamanders have long been rumored from the nearby Trinity Alps area. The most detailed report came from a deer-hunting attorney who claimed he saw five such animals in the New River in the 1920s, ranging up to an estimated nine feet long. Biologist Thomas Rodgers, who investigated this story, allowed it was possible that a relict population of giant salamanders still lived in California.

Rodgers also examined the Sacramento River specimen. The local press related a story that this salamander was an escapee named "Benny," from a shipment of exotic pets from "somewhere in China." With the question of the California salamander thus properly muddled, Rodgers led an expedition into the Trinity Alps in 1961 to search for the alleged amphibians. Rodgers' group found only well-known native salamanders under a foot in length, and he came away doubting any giant salamanders existed.

Other reports have trickled in since then, but no one has produced another giant salamander. We are left with the usual problems of one-specimen cases. Was the animal an exotic escapee? A last relic of a lost race? A chance specimen of a still-living population? The answer is always the same: we may never know.

The Elusive Long-Snouted Moth

Entomologist Gene Kritsky would like to discover a new moth. Not just any moth, but a particular kind. One that he knows exists—or existed—even though no one has ever seen it and there is no physical evidence for its existence.

The explanation for this odd-sounding paragraph offers an insight into scientific detective work.

Madagascar is home to a unique orchid, a flower known as *Angraecum longicalcar* which boasts a nectar tube sixteen inches deep. Orchids of this kind are pollinated by moths, which drink the liquid at the bottom of the nectar tube with a proboscis which must be nearly as long as the tube itself.

Accordingly, Dr. Kritsky postulates a moth with a fifteen-inch nectar tube. He estimates its wingspan at about six inches. The orchid is, to entomologists, ample proof or its existence. "The orchid and its pollinator must have co-evolved," Kritsky reasons. "The orchid could only survive if it had a moth pollinator." Unfortunately, neither Kritsky nor anyone else has been able to find the moth so far.

There is a precedent for this kind of reasoning. In 1862, Charles Darwin studied another orchid from Madagascar, this one with a twelve-inch nectar tube, and predicted the existence of a moth with an appropriately long proboscis. This species, a type of sphinx moth, was finally discovered in 1903. The insect keeps its long feeding apparatus (a highly specialized tongue) curled up until it lights on the flower and goes to work.

So where is Kritsky's even more impressively endowed moth? It could be extinct, a distinct possibility since the orchid itself hasn't been seen in years. Did something happen to the moths, so the flowers that depended on them died out? Or did collecting and development wipe out the orchid, leaving the moths to starve?

Hopefully, neither of these things has happened, and one day Dr. Kritsky will add a most unusual moth to the textbooks.

More Maybe Animals

There are undoubtedly more undiscovered animal species in the world than the few we've looked at here. From around the globe, there are tales of believable animals whose existence is not proven, but which do seem worthy of investigation.

Anthropologist George Agonino was in the U.S. Army in World War II, stationed in New Guinea, when he went fishing using the time-honored if unsportsmanlike military method of dropping a hand grenade into Lake Sentani. Among the stunned fish that drifted to the surface of the freshwater lake was a shark about twelve feet long. Agogino had only time to sketch the creature before it sank again. The only known freshwater shark is a bull shark which has taken up residence in Lake Nicaragua in Central America. That lake is not completely landlocked, and at least some of these adaptable sharks go back and forth to the sea.

Also from fresh water, Chinese biologist Xiang Lihao made a startling claim in 1985. He reported that he and several of his Xinjiang University students were surveying China's Lake Hanas when they saw and photographed several enormous reddish, salmonlike fish. At an estimated thirty-three feet long, the Lake Hanas giant would be the largest freshwater fish in the world. As of this writing, the alleged photographs have not been published.

According to herpetologist Wilfred Neill, fragmentary remains obtained from Celebes in 1954 appear to represent an undescribed species of freshwater crocodile. There are also reports of a small freshwater crocodile from New Britain, which Neill suggests is a relative of the New Guinea crocodile.

Zoologist/writer Ivan Sanderson reported encountering an enormous bat in the Assumbo Mountains of western Africa in 1932. He insisted there could be no mistake, because the animal startled him half to death by swooping directly at him at head level. Sanderson and naturalist Gerald Russell shot at the bat but failed to down it. They agreed the bat's wingspan was at least twelve feet, roughly twice the figure for the largest bat known to science. There are reports of something similar from widely scattered locations across the continent, although there is still no hard evidence.

Sanderson was a rather eccentric scientist, but his four decades of animal collecting in every nook and cranny of the globe gave him unsurpassed field experience. Accordingly, it's not surprising that his writings mention some other curiosities.

An intriguing episode was recorded from the same expedition that produced the giant bat. When a local tribesman in Cameroon sold the visitors some fish, members of a rival tribe scoffed at the man's catch and assured Sanderson they could bring in something much better.

The next day, a dozen straining men dragged ashore a gigantic stingray, over eleven feet long counting the tail. The stinger alone was nineteen inches long. Sanderson's mind protested that a gargantuan of this type had no business living in fresh water, but there it was. Unfortunately, he had no way to save such a specimen, and his account of the incident makes no attempt to identify the species, so it's unclear whether this mighty fish was a very lost oceanic type or something unique.

In 1940, while in Mexico's rural state of Nayarit, Sanderson purchased two skins of a most peculiar cat at a remote village market. The larger skin was about six feet long, not counting the tail, and indicated an animal with a short muzzle and long legs. The flanks were striped light and dark brown, and long hairs formed a prominent ruff around the neck.

The skins of this "ruffed cat," tragically, were lost in a flood. Sanderson saw one more for sale, but couldn't afford it, and no more have turned up since. The animal apparently resembled the still-enigmatic onza in build, but the coloration and the ruff mark it as something even stranger.

Finally, Sanderson and Russell figure in a most controversial episode. Reports of a large, mostly aquatic reptile have been filtering out of Central Africa for more than two centuries. This subject has been sensationalized by speculation about "living dinosaurs." Still, *something* has been reported by a scientist—more than once, in fact—so the case merits inclusion here.

Sanderson and his party were traveling down the Mainyu River in Cameroon in 1932 when they heard a "gargantuan gurgling roar," followed by the brief surfacing and submergence of a large black head, "like a seal's but flattened." The head was too large to belong to a crocodile or hippo.

Most recent reports come from the Likouala region of the People's Republic of the Congo, southeast of the Mainyu. American biologist Roy Mackal led expeditions into this area in 1980 and

1981. Anyone who believes Africa is entirely developed and/or explored should read the chronicle of Mackal's efforts. Struggling for days through untracked wilderness where most of the scattered inhabitants had never seen a Caucasian, Mackal's party collected many reports but found nothing more substantial than an unexplained trail of broken brush and now-indistinct plate-sized footprints.

In the same region, Congolese biologist Marcellin Agnagna in 1983 investigated reports of a strange animal in a large but shallow body of water named Lake Tele. Agnagna reported seeing a long-necked creature, with a visible length of about sixteen feet, swimming or wading away.

If there is something real here, there's no rush to tag it a "dinosaur." An unknown reptile could be a crocodilian (although the reported long neck weighs against this) or a monitor lizard. The largest known monitor, the Komodo dragon, has been measured at ten feet long and has been alleged to reach twice that, and it's quite at home in the water. Interestingly, Danish scientist Jorgen Birket-Smith did report seeing an unknown monitor, a sand-colored reptile at least seven feet long, in 1950 in what was then the French Cameroons.

Colonel William Hichens, a British colonial official in Tanzania in the 1930s, wrote about his fruitless pursuit of an alleged cat the native inhabitants called *mngwa* or *nunda*. This lion-sized, brindled gray feline was blamed for killing several people, including two policemen from whose bodies Hichens collected unidentifiable gray hairs. He sent these for expert analysis, but received only a rather unhelpful report concluding they were "probably cat."

Hichens admitted tales of strange animals "lose nothing in the telling," but argued that "a stretcher-load of clawed, mauled, and mangled man dumped at one's tent-door is no myth at all." English hunter Patrick Bowen also tracked the mngwa. He described its spoor as resembling "a leopard print as large as the largest lion's."

After about a month, the killings in this episode simply stopped. There have been other reports like this from East Africa, but no one has caught any animal, known or unknown, that might be behind them.

Another large predator, a striped, pantherlike marsupial, has been reported from the rugged forests of northern Queensland in Australia. Zoologist Maurice Burton collected the evidence for this "tiger-cat" in a 1952 article in the journal *Oryx*. He listed sightings by European immigrants going back to 1871, including naturalist George Sharp's report of a distinctly striped animal "larger and darker" than the thylacine. Burton concluded the creature "may yet appear in the flesh," although he thought the reports might concern the last remnants of an extinct or nearly extinct species. Australia did boast a prehistoric marsupial "lion" called *thylacoleo*.

A single antelope skin from Liberia, dating to before World War II, is preserved in the Zoological Museum of Berlin University. The animal in question is small by antelope standards and spotted all over. There have been no more specimens.

Another antelope, a gazelle known from its type specimen only, is in the same state of limbo. A large gazelle with a dark reddish gray coat was collected in the 1830s, allegedly from the Farasan Islands in the Red Sea. It hasn't been seen there or anywhere else since then. An undescribed type of gazelle is also reported from Forrur Island in the Persian Gulf.

Also in 1992, a British expedition to a remote valley in western Nepal returned with photographs of two very odd elephants. The beasts were over eleven feet tall, larger than any Asian elephants ever measured, and had weird-looking heads with high domes on top and prominent nasal bridges above their trunks. These latter features are associated with extinct prehistoric forms called stegodonts. Freak specimens, or living fossils?

Zoologist Bernard Heuvelmans has observed an unidentified wildcat, weighing about twenty pounds, on Ile du Levant in the Mediterranean. He suggests this feline is a variant of the African wild cat, which looks like a domestic tabby on steroids.

To end a book like this without at least mentioning such famous but unproven "maybe" animals as the yeti, the sasquatch, and the alleged denizens of Loch Ness would be closed-minded. Suffice to say that most zoologists continue to doubt the existence of these animals...but would love to be proven wrong.

Afterword

The International Society of Cryptozoology is a group of 900 scientists and interested laypeople dedicated to the discovery of new species and the rediscovery of those species verging on extinction or presumed extinct. The Society gives the study of such animals scientific credibility and disseminates information through its journal, *Cryptozoology,* the quarterly *ISC Newsletter,* and annual conferences. One point the ISC emphasizes is that the organization is interested in all animals from insects on up, not just the "monsters." The Society takes no position on the existence of any particular alleged animal. Contact: J. Richard Greenwell, Secretary, P.O. Box 43070, Tucson, AZ 85733.

Finally, this author publishes his own bimonthly newsletter, *Exotic Zoology,* to help readers keep abreast of the latest developments in the field. Interested readers of this book are invited to send comments, clippings, reports, or requests for subscription information to 3405 Windjammer Drive, Colorado Springs, CO 80920, or e-mail MattWriter@AOL.com.

References

Note on Classification

Gould, Stephen J. 1992. "What Is a Species?" *Discover,* December.

Wilson, Edward O. 1992. *The Diversity of Life.* Cambridge, MA: Belknap Press.

Section I

The Newest Animals

Anonymous. 1975. "Ghost from the Ice Age," *Science Digest,* November.

Anonymous. 1991. "Shy Leviathan," *Discover,* November.

Bright, Michael. 1989. *There are Giants in the Sea.* London: Robson Books.

Diamond, Jared. 1985. "How many unknown species are yet to be discovered?" *Nature,* June 13.

Homewood, Katherine M., and Alan Rodgers, 1985. "Tanzania's Newest Primate," *Animal Kingdom,* Vol. 88, No. 5.

Jouventin, P., and J-P Roux. 1983. "Discovery of a new albatross," *Nature,* September 15.

Ley, Willy. 1959. *Willy Ley's Exotic Zoology.* New York: Viking Press.

Pearce, Chris. 1978. "The oldest cat on Earth?" *International Wildlife,* July/August.

Raeburn, Paul. 1990. "Found: a new species," *Associated Press,* June 21.

World Wildlife Fund. 1992. "WWF Team Discovers New Species in Lost Corner of Vietnam," press release, July 16.

Beaked Whales

Anonymous. 1991. "Shy Leviathan," *Discover,* November.

Baker, Mary L. 1987. *Whales, Dolphins, and Porpoises of the World.* Garden City, New York: Doubleday & Co.

Ellis, Richard. 1980. *The Book of Whales.* New York: Alfred A. Knopf.

Ichihara, Tadayoshi, and Joseph Moore. 1966. In Norris, K.S. (ed.) *Whales, Dolphins, and Porpoises.* Berkeley, CA: U. of California Press.

Jensen, Albert C. 1979. *Wildlife of the Oceans.* New York: Harry N. Abrams, Inc.

Leatherwood, Stephen, and Randall R. Reeves. 1983. *The Sierra Club Handbook of Whales and Dolphins.* San Francisco: Sierra Club Books.

Mead, James G., and Roger S. Payne. 1975. "A Specimen of the Tasman Beaked Whale, *Tasmacetus Shepherdi*, From Argentina," *Journal of Mammology,* February.

Nowack, Ronald M. 1991. *Walker's Mammals of the World*. Baltimore: Johns Hopkins University Press.

Ralls, Katherine, and Robert L. Brownell, Jr. 1991. "A whale of a new species," *Nature*, April 18.

Watkins, W. A. 1976. "A Probable Sighting of a live *Tasmacetus Shepherdi* in New Zealand Waters," *Journal of Mammology*, May.

Watson, Lyall. 1981. *Sea Guide to Whales of the World*. New York: E.P. Dutton.

Yameneko

Anonymous. 1984. "Iriomote cat down to 50," *Oryx*, July.

Anonymous. 1975. "Iriomote cat survey finds a new pig," *BBC Wildlife*, February.

Guggisberg, C.A.W. 1975. *Wild Cats of the World*. New York: Taplinger Publishing.

Parker, Sybil P., ed. 1990. *Grzimek's Encyclopedia: Mammals*. New York: McGraw-Hill.

Pearce, Chris. 1978. "The oldest cat on Earth?" *International Wildlife*, July/August.

Shuker, Karl P. N. 1993. *The Lost Ark*. London: HarperCollins.

Megamouth

Anonymous. 1991. "Megamouth Reveals a Phantom Shark's Realm," *National Geographic*, March.

Castro, Jose I. 1983. *The Sharks of North American Waters*. College Station, TX: Texas A&M.

Doubilet, David. 1990. "Suruga Bay: In the Shadow of Mount Fuji," *National Geographic*, October.

Editors of Reader's Digest. 1986. *Sharks: Silent Hunters of the Deep*. New York: Reader's Digest Books.

Ellis, Richard. 1983. *The Book of Sharks*. New York: Alfred A. Knopf.

FitzGerald, Lisa M. 1991. "Megamouth Alive!," *Sea Frontiers*, February.

Hubbs, Carl L., and W. I. Follett. 1947. "*Lamna Ditropis*, New Species, the Salmon Shark of the North Pacific," *Copeia*, September 12.

Lineaweaver, Thomas H., and Richard Backus. 1970. *The Natural History of Sharks*. Philadelphia: J. B. Lippincott.

McCormick, Harold W., *et. al.* 1978. *Shadows in the Sea: the Sharks, Skates, and Rays*. New York: Stein and Day.

Springer, Victor G., and Joy P. Gold. 1989. *Sharks in Question*. Washington, D.C.: Smithsonian Institution Press.

Steel, Rodney. 1985. *Sharks of the World*. New York: Facts on File Publications.

Coelacanth

Anonymous. 1953. "Capture 'Living Fossil' Fish," *Science News Letter*, January 17.

Greenwell, J. Richard. 1994. "Prehistoric fishing," *BBC Wildlife*, March.

Ley, Willy. 1959. *Willy Ley's Exotic Zoology*. New York: Viking Press.

Thomson, Keith S. 1991. *Living Fossil: The Story of the Coelacanth*. New York: W.W. Norton.

Itundu

Ley, Willy. 1959. *Willy Ley's Exotic Zoology.* New York: Viking Press.

Matthiessen, Peter. 1991. *African Silences.* New York: Random House.

Mountfort, Guy. 1988. *Rare Birds of the World.* London: William Collins & Sons.

Traveling Ants

Anonymous. 1991. "Office Lunches Draw a New Ant Species," *National Geographic,* April.

Anonymous. 1982. "No Taste of Honey for Flesh-Eating Bees," *New Scientist,* August 19.

Gain, Bruce. 1993. "Cave Dwellers," *Popular Science,* January.

Wilson, Edward O. 1992. *The Diversity of Life.* Cambridge, MA: Belknap Press.

Wood, Gerald L. 1977. *Animal Facts and Feats.* Sterling Publishing Co., New York.

New Parrots

Anonymous. 1988. "Another Polly," *Science Digest,* August.

O'Neill, John P., *et. al.* 1991. "*Nannopsittaca Dachilleae,* A New Species of Parrotlet From Eastern Peru," *Auk,* Vol. 108, No. 2.

Lowery, George H., Jr., and John P. O'Neill. 1969. "A New Species of Antipatta From Peru and a Revision of the Subfamily *Grallariinae,*" *Auk,* January.

Lowery, George H., Jr., and John P. O'Neill. 1964. "A New Genus and Species of Tanager From Peru," *Auk,* April.

Ridgely, Robert. 1992. Personal communication, June 19.

Stap, Don. 1990. *A Parrot Without a Name.* New York: Knopf.

Stap, Don. 1985. "The Bird on the Right is a Yellow-Bellied Mezzo-Soprano," *International Wildlife,* July-August.

Archey's Frog

Cherfas, Jeremy. 1984. "Ulcer studies rescued by reincarnated frog," *BBC Wildlife,* April.

Grzimek, Bernhard (ed.) 1979. *Grzimek's Animal Encyclopedia.* Vol. 5. New York: Van Nostrand.

Kanze, Edward. 1992. *Notes From New Zealand.* New York: Henry Holt & Co.

Kingdon, Jonathan. 1989. *Island Africa.* Princeton, NJ: Princeton University Press.

Platz, James E. 1993. "*Rana subaquavocalis,* a Remarkable New Species of Leopard Frog (*Rana pipiens* Complex) from Southeastern Arizona that Calls Under Water," *Journal of Herpetology,* Vol. 27, No. 2.

Kouprey

Anonymous. 1992. "WCU's Species Survival Commission Surveys Future of the Kouprey." *International Wildlife,* September/ October.

Bright, Michael. 1987. *The Living World.* New York: St. Martin's Press.

Burton, John A., and Bruce Pearson. 1987. *The Collins Guide to the Rare Mammals of the World.* Lexington, MA: The Stephen Greene Press.

Shuker, Karl P. N. 1993. *The Lost Ark.* London: HarperCollins.

Thouless, Chris. 1987. "Kampuchean wildlife - survival against the odds," *Oryx,* October.

Chacoan Peccary

Anonymous. 1975. "Ghost from the Ice Age," *Science Digest,* November.

Mayer, John, and Ralph Wetzel. 1986. "*Catagonus wagneri,*" *Mammalian Species,* no. 259. The American Society of Mammologists, June 16.

Wetzel, Ralph M., *et. al.* 1975. "*Catagonus,* an 'Extinct' Peccary, Alive in Paraguay," *Science,* August 1.

Vaquita

Brown, Martha. 1987. "Searching for the Vaquita," *Defenders,* May-June.

Mulvaney, Kevin, *et. al.* 1990. *The Greenpeace Book of Dolphins.* New York: Sterling Publishing Company.

Undersea Worlds

Anonymous. 1982. "Eellike Fish Discovered on Ocean Floor." *Science Digest,* May.

Anonymous. 1990. "Lake Baikal's Deep Vent: A Freshwater First," *National Geographic,* December.

Alper, Joseph. 1990. "Methane Eaters," *Sea Frontiers,* December.

Bright, Michael. 1989. *There are Giants in the Sea.* London: Robson Books.

Brownlee, Shannon. 1984. "Bizarre Beasts of the Abyss," *Discover,* July.

Cone, Joseph. 1991. *Fire Under the Sea.* New York: William Morrow and Company.

Friend, Tim. 1993. "Life forms thrive at the volatile edge," *USA Today,* May 4.

Grassle, J. Frederick. 1985. "Hydrothermal Vent Animals: Distribution and Biology," *Science,* August 23.

Jannasch, Holger W., and Michael J. Mottl. 1985. "Geomicro-biology of Deep-Sea Hydrothermal Vents," *Science,* August 23.

Lutz. Richard A., and Rachel M. Haymon. 1994. "Rebirth of a Deep-sea Vent," *National Geographic,* November.

Rona, Peter A. 1992. "Deep-Sea Geysers of the Atlantic," *National Geographic,* October.

Vu Quang Ox

Adler, Hans J. 1995. "Antelope Expose," *BBC Wildlife,* January.

Anonymous. 1994. "No Bull," *USA Today,* July 24.

Anonymous. 1993. "Vietnamese Animals - Lost and Found," *Popular Science,* September.

Anonymous. 1992. "Early Whatizit is found," *USA Today,* August 17.

Anonymous. 1992. "No Ordinary Ungulate," *Science,* August 7.

Anonymous. 1989. "New species of deer in China," *Oryx,* April.

Anonymous. 1987. "New gazelle," *Oryx,* January.

Burton, John A., and Bruce Pearson. 1987. *The Collins Guide to the Rare Mammals of the World.* Lexington, MA: The Stephen Greene Press.

Connor, Steve. 1994. "Lost worlds rich in unique wildlife," *The Independent on Sunday,* July 3, pp. 12-13.

East, Rod. 1992. "Conservation Status of Antelopes in Asia and the Middle East, Part 1," *Species,* December.

Groves, Colin P., and Douglas M. Lay. 1985. "A new species of the genus Gazella from the Arabian Peninsula," *Mammalia,* Vol. 49, No. 1.

Linden, Eugene. 1994. "Ancient Creatures in a Lost World," *TIME,* June 20, pp. 52-4.

Nowack, Ronald M. 1991. *Walker's Mammals of the World.* Baltimore: Johns Hopkins University Press.

Rabinowitz, Alan. 1994. Personal communication, August 31.

Reynolds, Clarence V. 1994. "A New Cow," *Discover,* January.

Schaller, George B. 1995. "An Unfamiliar Bark," *Wildlife Conservation,* June.

Scott, Kate. 1994. "Vietnam Explosion," *BBC Wildlife,* June.

Van Dung, Vu, with Pham Mong Giao, Nguyen Ngoc Chinh, Do Tuoc, Peter Arctander and John MacKinnon. 1993. "A new species of living bovid from Vietnam," *Nature,* June 3, p. 443.

Shuker, Karl. 1995. "Vietnam - why scientists are stunned," *Wild About Animals,* March.

Wang, Ma Shilai, and Yingxiang Shi Liming. 1990. "A New Species of the Genus *Muntiacus* From Yunnan, China," *Acta Zoologica Sinica,* p.52.

World Wide Fund for Nature. 1994. "Vu Quang Ox Found - Live!" Press release, June 24.

World Wide Fund for Nature. 1994. "Scientists Find Second New Mammal in Vietnam," press release, April 21.

World Wide Fund for Nature. 1993. "Additional Evidence Confirms New Vietnam Species," press release, March 29.

World Wide Fund for Nature. 1992. "WWF Team Discovers New Species in Lost Corner of Vietnam," press release, July 16.

Neopilina

Batten, Roger L. 1984. "*Neopilina, Neomphalus* and *Neritopsis,* Living Fossil Molluscs," in Eldredge, Niles, and Steven M. Stanley (eds). *Living Fossils.* New York: Springer Verlag.

Bright, Michael. 1987. *The Living World.* New York: St. Martin's Press.

Cromie, William J. 1966. *The Living World of the Sea.* Englewood Cliffs, NJ: Prentice-Hall.

Huyghe, Patrick. 1993. "New-Species Fever," *Audubon,* March-April.

Kaharl, Victoria A. 1990. *Water Baby: the Story of Alvin.* New York: Oxford University Press.

Lewin, Roger. 1983. "New Phylum Discovered, Named," *Science,* October 14.

Lipske, Mike. 1990. "Wonder Holes," *International Wildlife,* January-February.

Soule, Gardner (ed.) 1968. *Under the Sea.* New York: Meredith Press.

Svitil, Kathy. 1993. "It's Alive, and It's a Graptolite," *Discover,* July.

Taylor, Mike. 1993. "Home and Away," *BBC Wildlife,* March.

Wilson, Craig. 1994. "Presiding over a shrine to seashells," *USA Today*, September 28.

Wilson, Edward O. 1992. *The Diversity of Life.* Cambridge, Mass.: Belknap Press.

Tamarin and marmoset

Anonymous. 1993. "Welcome to the Order," *Discover,* March.

Anonymous. 1991. "Stunning New Primate Species Found in Brazil," *National Geographic,* October.

Anonymous. 1989. "Scientists Identify a New Lemur Species," *National Geographic,* October.

Burton, John A., and Bruce Pearson. 1987. *The Collins Guide to the Rare Mammals of the World.* Lexington, MA: The Stephen Greene Press.

Homewood, Katherine, with W. Alan Rodgers. 1985. "Tanzania's Newest Primate," *Animal Kingdom,* September/October.

Huyghe, Patrick. 1993. "New-Species Fever," *Audubon,* March-April.

Mallinson, Jeremy. 1989. *Travels in Search of Endangered Species.* London: David and Charles.

Maranto, Gina. 1986. "Will Guenons Make a Monkey of Darwin?" *Discover,* November.

Nowack, Ronald M. 1991. *Walker's Mammals of the World.* Baltimore: Johns Hopkins University Press.

Raeburn, Paul. 1992. "Remote Amazon gives old world new monkey," Associated Press, October 12.

Reilly, Patti. 1992. "500 Years Later, Scientists Continue to 'Discover' America: New Primate Species Found in Brazil," news release, Conservation International. Washington, D.C.

Schaller, George B. 1990. "Saving China's Wildlife," *International Wildlife,* January-February.

Jellyfish

Anderson, Robert. 1992. "High Seas Drifter," *Natural History,* November.

Boletzky, S. V., et. al. 1992. "Octopod 'ballooning' response," *Nature,* March 19.

Bright, Michael. 1989. *There are Giants in the Sea.* London: Robson Books.

Bright, Michael. 1994. Personal communication, February 4.

Friend, Tim. 1994. "Same-sex tendency, or rather, tentacles," *USA Today,* October 13.

Friend, Tim. 1994. "Groundswell of research deep in Monterey Bay," *USA Today,* February 22.

Heuvelmans, Bernard. 1986. "Annotated Checklist of Apparently Unknown Animals With Which Cryptozoology is Concerned," *Cryptozoology,* Vol. 5.

Kaharl, Victoria A. 1990. *Water Baby: the Story of Alvin.* New York: Oxford University Press.

Raynal, Michel. 1987. "The Linnaeus of the Zoology of Tomorrow," *Cryptozoology,* Vol. 6.

Parade of Mammals

Anonymous. 1990. "Rare Animal Finds Excite Scientists," *International Wildlife,* July-August.

Burton, John A., and Bruce Pearson. 1987. *The Collins Guide to the Rare Mammals of the World.* Lexington, MA: The Stephen Greene Press.

Champkin, Julian. 1994. "Out of the trees, the kangaroo family's new branch," *Daily Mail,* July 21, p.19.

Domico, Terry. 1993. *Kangaroos: The Marvelous Mob.* New York: Facts on File.

Henry, Stephen. 1990. "Things That Go Swoop in the Night," *International Wildlife,* September-October.

Izor, Robert J., and Luis de la Torre. 1978. "A New Species of Weasel (*Mustela*) from the Highlands of Columbia," *Journal of Mammology,* February.

Nowack, Ronald M. 1991. *Walker's Mammals of the World.* Baltimore: Johns Hopkins University Press.

Pine, Ronald H. 1972. "A New Subgenus and Species of Murine Opossum (Genus Marmosa) From Peru," *Journal of Mammology,* May.

Pritchard, J. S. 1989. "Ilin Island cloud rat extinct?" *Oryx,* July.

Shuker, Karl P.N. 1989. *Extraordinary Animals Worldwide.* London: Robert Hale.

Thornback, Jane, and Martin Davis, compilers. 1982. *The IUCN Mammal Red Data Book.* Gland, Switzerland: International Union for the Conservation of Nature.

Van Dyck, Stephen. 1987. "The Bronze Quoll, *Daysurus spartacus* (Marsupalia: Dasyuridae), a New Species from the Savannahs of Papua New Guinea," *Australian Mammology*, No. 2.

Galathea Voyage

Bruun, Anton F., *et. al.*, editors. 1956. *The Galathea Deep Sea Expedition.* Translated by Reginald Spink. New York: The MacMillan Company.

Kitti's Hog-Nosed Bat

Nowack, Ronald M. 1991. *Walker's Mammals of the World.* Baltimore: Johns Hopkins University Press.

McNeely, Jeffrey A., and Paul S. Wachtel. 1988. *Soul of the Tiger.* New York: Athena Books.

Wood, Gerald L. 1977. *Animal Facts and Feats.* Sterling Publishing Co., New York.

Smallmouth and Longfin Char

Berra, Tim M. and Rita M. Berra. 1977. "A Temporal and Geographical Analysis of New Teleost Names Proposed at 25 Year Intervals from 1869-1970," *Copeia,* No.4.

Choen, Daniel M., and Bennie A. Rohr. 1993. "Description of a Giant Circumglobal *Lamprogrammus* Species (Pisces: Ophidiidae)," *Copeia,* No. 2.

Shuker, Karl P. N. 1993. *The Lost Ark.* London: HarperCollins.

Skopets, Mikhail. 1992. "Secrets of Siberia's White Lake," *Natural History,* November.

Crested Iguana

Anonymous. 1984. "New turtle in Brazil," *Oryx,* April.

Bustard, Robert. 1972. *Sea Turtles.* New York: Taplinger Publishing.

Gibbons, John R.H. 1984. "Discovery of a Brand-New Million-Year-Old Iguana," *Animal Kingdom,* February-March.

Gibbons, John R.H. 1984. "On the Trail of the Crested Iguana," *Animal Kingdom,* December/January.

Greenwell, J. Richard. 1990. "Yemen Monitor Described," *ISC Newsletter*, Autumn. (Quoting Bohme, Wolfgang, with Ulrich Joger and Beat Schatti. 1989. "A New Monitor Lizard [Reptilia: Varanidae] from Yemen, with Notes on Ecology, Phylogeny and Zoogeography, *Fauna of Saudi Arabia,* vol. 10.)

Groombridge, Brian, compiler. 1982. *The IUCN Amphibia-Reptilia Red Data Book*. Gland, Switzerland: International Union for the Conservation of Nature.

Huyghe, Patrick. 1993. "New-Species Fever," *Audubon*, March-April.

Neill, Wilfred T. 1971. *The Last of the Ruling Reptiles*. New York: Columbia University Press.

Tennesen, Michael. 1985. "Crawling Out of Limbo," *International Wildlife*, July-August.

Rufous-Winged Sunbird

Ehrlich, Paul, David Dobkin, and Darryl Wheye. 1992. *Birds in Jeopardy*. Stanford, CA: Stanford University Press.

Fuller, Errol. 1987. *Extinct Birds*. New York: Facts on File.

Huyghe, Patrick. 1993. "New-Species Fever," *Audubon*, March-April.

Jensen, Fleming P. 1985. "Something New Under the Sun," *Animal Kingdom*, September/October.

Jouventin, P., and J-P Roux. 1983. "Discovery of a new albatross," *Nature*, September 15.

Mountfort, Guy. 1988. *Rare Birds of the World*. Lexington, MA: Stephen Greene Press.

Shuker, Karl P. N. 1993. *The Lost Ark*. London: HarperCollins.

Zimmer, John T., and Ernst Mayr. 1943. "New Species of Birds Described from 1938 to 1941," *Auk*, April.

Section II

Animals Lost and Found

Anonymous. 1989. "Serpent eagle rediscovered on FFPS-funded expedition," *Oryx*, April.

Anonymous. 1992. "Draft IUCN Categories of Threat for Species," *Species*, December.

Caras, Roger A. 1966. *Last Chance on Earth*. Philadelphia: Chilton Books.

Lowe, David, et al. 1990. *The Official World Wildlife Fund Guide to Endangered Species of North America*. Washington, D.C.: Beacham Publishing.

Mackal, Roy. 1980. *Searching for Hidden Animals*. New York: Doubleday.

Roland, Ardith E. 1992. "Tonkin Snub-nosed Monkey Rediscovered," *Species*, December.

Steller's Sea Cow

Dietz, Tim. 1992. *The Call of the Siren*. Golden, CO: Fulcrum Publishing.

Haley, Delphine. 1978. "The Saga of Steller's Sea Cow," *Natural History*, November.

Heuvelmans, Bernard. 1968. *In the Wake of the Sea-Serpents*. New York: Hill and Wang.

Mackal, Roy. 1980. *Searching for Hidden Animals*. New York: Doubleday.

Parker, Sybil B. (ed.). 1990. *Grzimek's Encyclopedia: Mammals*. New York: McGraw-Hill.

Stejneger, Leonhard. 1936. *Georg Wilhelm Steller*. Cambridge, MA: Harvard University Press.

Thylacine

Bunk, Steve. 1985. "Just How Extinct is Tasmania's Tiger?" *International Wildlife,* July/August.

Douglas, Athol M. 1986. "Tigers in Western Australia," *New Scientist,* April 24.

Guiler, Eric R. 1985. *Thylacine: The Tragedy of the Tasmanian Tiger.* Oxford: Oxford University Press.

Guiler, Eric R. 1966. "In pursuit of the Thylacine," *Oryx.*

Matthews, Peter (ed.). 1993. *The Guinness Book of Records.* New York: Facts on File.

Sayles, Jim. 1979. "Stalking the Tasmanian Tiger," *Animal Kingdom,* December/January.

Sutherland, Struan. 1995. "Spotting puts new bite into tiger tale," *Australian Doctor,* March 31.

Takahe

Anonymous. 1977. "Takahe Rallies," *Oryx.* December.

Caras, Roger A. 1966. *Last Chance on Earth.* Philadelphia: Chilton Books.

Fuller, Errol. 1987. *Extinct Birds.* New York: Facts on File.

Guiffre, Philippe. 1993. "Le choc du moa," *Le Pointe,* February 6.

Kanze, Edward. 1992. *Notes From New Zealand.* New York: Henry Holt & Co.

Various. 1995. New Zealand newspaper clippings, January 25th to February 14. (Provided by Raymond Nelke.)

Eskimo Curlew

Anonymous. 1988. "Eyes on the Eskimo Curlew," *Natural History,* April.

Bodsworth, Fred. 1987. *Last of the Curlews.* New York: Dodd, Mead & Co.

Caras, Roger A. 1966. *Last Chance on Earth.* Philadelphia: Chilton Books.

Lowe, David, et al. 1990. *The Official World Wildlife Fund Guide to Endangered Species of North America.* Washington, D.C.: Beacham.

Eastern Cougar

Anonymous. *USA Today,* 1994. September 16.

Allen, Thomas B. 1974. *Vanishing Wildlife of North America.* Washington, D.C.: National Geographic.

Downing, Robert L. 1982. *Eastern Cougar Recovery Plan.* Atlanta, GA: U.S. Fish and Wildlife Service.

Lowe, David, et al. 1990. *The Official World Wildlife Fund Guide to Endangered Species of North America.* Washington, D.C.: Beacham.

Lutz, John A. 1995. "*1994 Statistical Review of Felis concolor Sightings.*" Baltimore: Eastern Puma Research Network.

McNamee, Thomas. 1980. "Chasing a Ghost," *Audubon,* March.

Thornbeck, Jane, and Martin Jenkins, compilers. 1982. *The IUCN Mammal Red Data Book.* Gland, Switzerland: International Union for the Conservation of Nature.

Tinsley, Jim Bob. 1987. *The Puma.* El Paso: Texas Western Press.

Wright, Bruce S. 1961. "The Latest Specimen of the Eastern Puma," *Journal of Mammology,* May.

Wright, Bruce S. 1960. "The Return of the Cougar," *Audubon,* November-December.

Survivor Seals

Bonner, Nigel. 1994. *Seals and Sea Lions of the World.* New York: Facts on File.

Caras, Roger A. 1966. *Last Chance on Earth.* Philadelphia: Chilton Books.

Curry-Lindahl, Kai. 1972. *Let Them Live.* New York: William Morrow & Co.

Rutherford, Donald. 1973. "Nearly, if Not Quite Extinct," *Sea Frontiers,* November/December.

Thornbeck, Jane, and Martin Jenkins, compilers. 1982. *The IUCN Mammal Red Data Book.* Gland, Switzerland: International Union for the Conservation of Nature.

Wood, Gerald L. 1977. *Animal Facts and Feats.* New York: Sterling Publishing Co.

Dusky Seaside Sparrow

Avise, John C., and William S. Nelson. 1989. "Molecular Genetic Relationships of the Extinct Dusky Seaside Sparrow," *Science,* February 3.

Bergman, Charles. 1990. *Wild Echoes.* New York: McGraw-Hill.

Cadieux, Charles L. 1991. *Wildlife Extinction.* Washington, D.C.: Stonewall Press.

Walters, Mark J. 1992. *A Shadow and a Song.* Post Mills, Vt. Chelsea Green Publishing Co.

Ivory-Billed Woodpecker

Anonymous. 1993. "Ivory-billed woodpecker extinct," *Oryx,* October.

Cadieux, Charles L. 1991. *Wildlife Extinction.* Washington, D.C.: Stonewall Press.

Caras, Roger A. 1966. *Last Chance on Earth.* Philadelphia: Chilton Books.

Lowe, David, et al. 1990. *The Official World Wildlife Fund Guide to Endangered Species of North America.* Washington, D.C.: Beacham Publishing.

Short, Lester, and Jennifer Horne. 1986. "The Ivorybill Still Lives," *Natural History.* July.

Short, Lester. 1993. *The Lives of Birds.* New York: Henry Holt and Company.

Shuker, Karl P. N. 1993. *The Lost Ark.* London: HarperCollins.

Sinai Leopard

Anonymous. 1975. "Back from extinction," *Wildlife.* May.

Arrighi, J., and M. Salotti. 1988. "Le chat sauvage (*Felis sylvestris* Schreber, 1777) en Corse. Confirmation de da presence et approche taxonomique," *Mammalia.*

Line, Les, and Edward R. Ricciuti. 1985. The *Audubon Society Book of Wild Cats.* New York: Harry N. Abrams, Inc.

Shuker, Karl P. N. 1989. *Mystery Cats of the World.* London: Robert Hale.

King Bee

Bright, Michael. 1987. *The Living World.* New York: St. Martin's Press.

Shuker, Karl P.N. 1989. *Extraordinary Animals Worldwide.* London: Robert Hale.

Huia

Bright, Michael. 1987. *The Living World.* New York: St. Martin's Press.

Mountfort, Guy. 1988. *Rare Birds of the World.* Lexington, Mass.: Stephen Greene Press.

Shuker, Karl P.N. 1989. *Extraordinary Animals Worldwide.* London: Robert Hale.

Stivens, Dal. 1976. "Has the Huia Gone for Good?" *BBC Wildlife,* July.

177

Tarpan

Balouet, Jean-Christophe, and Eric Alibert. 1990. *Extinct Species of the World.* New York: Barron's.

Cherfas, Jeremy. 1991. "Ancient DNA: Still Busy After Death," *Science,* September 20.

Day, David. 1990. *The Doomsday Book of Animals.* New York: Viking Press.

Groves, Colin P. 1974. *Horses, Asses, and Zebras in the Wild.* Hollywood, Fla.: Ralph Curtis Books.

McClung, Robert M. 1976. *Lost Wild Worlds.* New York: William Morrow.

Przewalski's Horse

Anonymous. 1992. "Wild Asian Horse to Return Home," *National Geographic,* March.

Burton, John A., and Bruce Pearson. 1987. *The Collins Guide to the Rare Mammals of the World.* Lexington, MA: The Stephen Greene Press.

Groves, Colin P. 1974. *Horses, Asses, and Zebras in the Wild.* Hollywood, Fla.: Ralph Curtis Books.

Stone, Richard. 1993. "Zoo Horse May Heed Call of the Wild," *Science,* July 2.

Tsevegmid, D., and A. Dashdorj. 1974. "Wild Horses and other Endangered Wildlife in Mongolia," *Oryx,* February.

Fraser's Dolphin

Burton, John A., and Bruce Pearson. 1987. *The Collins Guide to the Rare Mammals of the World.* Lexington, MA: The Stephen Greene Press.

Fraser, F.C. 1966. "Comments on the Delphinoidea." In Norris, K. S. (ed.). 1966. *Whales, Dolphins, and Porpoises.* Berkeley, CA: University of California Press.

Mulvaney, Kevin, et al. 1990. *The Greenpeace Book of Dolphins.* New York: Sterling Publishing Company.

Perrin, W.F., *et al.* 1981. "*Stenella Clymene,* a Rediscovered Tropical Dolphin of the Atlantic," *Journal of Mammology,* August.

Perrin, W.F., *et al.* 1973. "Rediscovery of Fraser's Dolphin *Lagenodelphis hosei,*" *Nature,* Vol. 241.

Santo Mountain Starling

Anonymous. 1992. "Photographic Debut for a Pacific Starling," *National Geographic,* June.

McClung, Robert M. 1993. *Lost Wild America.* Hamden, Connecticut: Linnet Books.

Mountfort, Guy. 1988. *Rare Birds of the World.* Lexington, MA: Stephen Greene Press.

Schreiber, Rudolf L., with Antony W. Diamond, Roger Tory Peterson, and Walter Cronkite. 1989. *Save the Birds.* Boston: Houghton Mifflin.

Passenger Pigeon and Carolina Parakeet

Buscemi, Doreen. 1978. "The Last American Parakeet," *Natural History,* April.

Ehrlinger, David. (Director of Horticulture, Cincinnati Zoo.) 1993. Personal communication, October 6.

Fuller, Errol. 1987. *Extinct Birds.* New York: Facts on File.

Hadley, Philip. 1930. "The Passenger Pigeon," *Science,* February 14.

Shuker, Karl P. N. 1989. *Extraordinary Animals Worldwide.* London: Robert Hale.

Quagga

Anonymous. 1992. "Can the Quagga Defy Extinction's Quagmire?" *National Geographic,* July.

d'Alessio, Vittorio. 1992. "Born-again Quagga Defies Extinction," *New Scientist,* November 30.

Groves, Colin P. 1974. *Horses, Asses, and Zebras in the Wild.* Hollywood, Fla.: Ralph Curtis Books.

Lemurs

Anonymous. 1990. "Rare Animal Finds Excite Scientists," *International Wildlife,* July/August.

Jolly, Alison. 1988. "Madagascar's Lemurs: On the Edge of Survival," *National Geographic,* August.

Aye-Aye

Anonymous. 1992. "Captive-Born Aye-Aye Delights Scientists," *International Wildlife,* July/August.

Anonymous. 1993. "Aye-aye captive breeding," *Oryx,* January.

Burbank, Jeffrey. 1992. "The Aye-Ayes Have It," *ZooLife,* Fall.

Tigers

Anonymous. 1980. "Bali may still have large cats," *BBC Wildlife,* February.

Anonymous. 1993. "Does the Javan Tiger Survive?" *Cat News,* IUCN Cat Specialist Group, March.

Becker, John. 1993. "ISEC Supports Javan Tiger Project," *Cat Tales,* Summer.

Curry-Lindahl, Kai. 1972. *Let Them Live.* New York: William Morrow & Co.

Day, David. 1990. *The Doomsday Book of Animals.* New York: Viking Press.

Jackson, Peter. 1993. "Status of the Tigers of the World," *CBSG News* (Captive Breeding Specialist Group, IUCN), December.

Parker, Sybil P. (Ed.) 1990. *Grzimek's Encyclopedia: Mammals.* New York: McGraw-Hill.

Seidensticker, John. 1987. "Bearing Witness: Observations on the Extinction of *Panthera tigris balica* and *Panthera tigris sondaica,*" in Tilson, Ronald L., and Ulysses S. Seal, (eds.) 1987. *Tigers of the World.* Park Ridge, New Jersey: Noyes Publications.

Shuker, Karl P. N. 1989. *Mystery Cats of the World.* London: Robert Hale.

Black-Footed Ferret

Anonymous. 1993. "Reintroduced ferrets breeding," *Oryx,* January.

Anonymous. 1991. "Black-footed Ferrets: a Risky Return to the Wild," *National Geographic,* September.

Bergman, Charles. 1990. *Wild Echoes.* New York: McGraw-Hill.

Godbey, Jerry, and Dean Biggins. 1994. "Recovery of the Black-footed Ferret: Looking Back, Looking Forward," *Endangered Species Technical Bulletin,* January/February.

Noreen, Patty. 1994. "Ferrets don't leave home without a fuss," *Colorado Springs Gazette-Telegraph,* September 16.

Mysterious Starling

Fuller, Errol. 1987. *Extinct Birds*. New York: Facts on File.

Cahow

Fisher, James, with Noel Simon and Jack Vincent. 1969. *Wildlife in Danger*. New York: Viking Press.

Fuller, Errol. 1987. *Extinct Birds*. New York: Facts on File.

Lever, Christopher. 1984. "Conservation success for two Bermudan bird species," *Oryx*, July.

Mountfort, Guy. 1988. *Rare Birds of the World*. Lexington, Mass.: Stephen Greene Press.

Schreiber, Rudolf L., with Antony W. Diamond, Roger Tory Peterson, and Walter Cronkite. 1989. *Save the Birds*. Boston: Houghton Mifflin.

Silverberg, Robert. 1967. *The Auk, the Dodo, and the Oryx*. New York: Thomas Y. Crowell Co.

Watling, Dick. 1987. "The Fiji Petrel: Stranger in Paradise," *Animal Kingdom*, January-February.

Pygmy Hog

Mallinson, Jeremy. 1989. *Travels in Search of Endangered Species*. London: David and Charles.

Parker, Sybil P. (Ed.) 1990. *Grzimek's Encyclopedia: Mammals*. New York: McGraw-Hill.

Bachman's Warbler

Lowe, David, et al. 1990. *The Official World Wildlife Fund Guide to Endangered Species of North America*. Washington, D.C.: Beacham Publishing.

Mountfort, Guy. 1988. *Rare Birds of the World*. Lexington, Mass.: Stephen Greene Press.

Schreiber, Rudolf L., with Antony W. Diamond, Roger Tory Peterson, and Walter Cronkite. 1989. *Save the Birds*. Boston: Houghton Mifflin.

Terborgh, John. 1989. *Where Have All the Birds Gone?* Princeton, N.J.: Princeton University Press.

Bandicoots and Wallaroos

Anonymous. 1994. "Marsupial Thought Extinct is Found," *Dayton* (Ohio) *Dispatch*, December 8.

Burton, John (ed.) 1991. *The Atlas of Endangered Species*. New York: MacMillan Publishing.

Burton, John A., and Bruce Pearson. 1987. *The Collins Guide to the Rare Mammals of the World*. Lexington, Mass.: The Stephen Greene Press.

Caras, Roger A. 1966. *Last Chance on Earth*. Philadelphia: Chilton Books.

Domico, Terry. 1993. *Kangaroos: The Marvelous Mob*. New York: Facts on File.

Parker, Sybil P. (Ed.) 1990. *Grzimek's Encyclopedia: Mammals*. New York: McGraw-Hill.

Wanted Birds

Anonymous. 1989. "Serpent eagle rediscovered on FFPS-funded expedition," *Oryx*, April.

Balouet, Jean-Christophe, and Eric Alibert. 1990. *Extinct Species of the World.* New York: Barron's.

Day, David. 1990. *The Doomsday Book of Animals.* New York: Viking.

Fitter, Richard. 1974. "25 Years On: a Look at Endangered Species," *Oryx,* February.

Fuller, Errol. 1987. *Extinct Birds.* New York: Facts on File.

Mountfort, Guy. 1988. *Rare Birds of the World.* London: William Collins & Sons.

Thin-Spined Porcupine

Anonymous. 1981. "*C. Subspinosus,* I presume?" *Science News,* Vol. 131.

Parker, Sybil P. (Ed.) 1990. *Grzimek's Encyclopedia: Mammals.* New York: McGraw-Hill.

Yoon, Carol. 1995. "Woolly Flying Squirrel, Long Thought Extinct, Shows Up in Pakistan," *New York Times,* March 14.

Shadowy Bats

Anonymous. 1993. "Bat rediscovery," *Oryx,* October.

Burton, John A., and Bruce Pearson. 1987. *The Collins Guide to the Rare Mammals of the World.* Lexington, Mass.: The Stephen Greene Press.

Nowack, Ronald M. 1991. *Walker's Mammals of the World.* Baltimore: Johns Hopkins University Press.

Galapagos Tortoise

Alberts, Alison. 1993. "The Most Endangered Lizard in the World: the Jamaican Iguana, *Cyclura collei,*" *The Vivarium,* July/August.

Balouet, Jean-Christophe, and Eric Alibert. 1990. *Extinct Species of the World.* New York: Barron's.

Pritchard, Peter. 1992. "Time Out for Turtles," *Wildlife Conservation,* July/August.

Jerdon's Courser and Pink-Headed Duck

Anonymous. 1986. "Jerdon's Courser Alive and Well and Living in India," *New Scientist,* April 3.

Fitter, Richard. 1974. "25 Years On: a Look at Endangered Species," *Oryx,* February.

Nugent, Rory. 1991. *The Search For the Pink-Headed Duck.* Boston: Houghton Mifflin Co.

Shuker, Karl P.N. 1989. *Extraordinary Animals Worldwide.* London: Robert Hale.

Section III

The Continuing Search

Goodwin, George C. 1946. "Inopinatus the Unexpected," *Natural History,* November.

Mackal, Roy. 1980. *Searching for Hidden Animals.* New York: Doubleday.

Wood, Gerald L. 1977. *Animal Facts and Feats.* New York: Sterling Publishing Co.

Onza

Alderton, David. 1993. *Wild Cats of the World.* New York: Facts on File.

Best, Troy. 1993. Personal communication, January 15.

James, Jamie. "Bigfoot or Bust," 1988. *Discover,* March.

Greenwell, J. Richard. 1992. Personal communication, July 7.

Marshall, Robert E. 1961. *The Onza.* New York: Exposition Press.

Tinsley, Jim Bob. 1987. *The Puma.* El Paso: Texas Western Press.

Steller's Sea Monkey

Bright, Michael. 1989. *There Are Giants in the Sea.* London: Robson Books.

Goodwin, George C. 1946. "The End of the Great Northern Sea Cow," *Natural History,* February.

Mackal, Roy. 1980. *Searching for Hidden Animals.* New York: Doubleday.

Stejneger, Leonhard. 1936. *Georg Wilhelm Steller.* Cambridge, MA: Harvard University Press.

Strangest Dolphin

Corliss, William R. 1970. *Mysteries Beneath the Sea.* New York: Thomas Y. Crowell Co.

Heuvelmans, Bernard. 1968. *In the Wake of the Sea-Serpents.* New York: Hill and Wang.

Raynal, Michel. 1991. "Cetaceans with two dorsal fins," *Aquatic Mammals,* Vol. 17.1.

Marozi

Gandar Dower, Kenneth C. 1937. *The Spotted Lion.* Boston: Little, Brown and Company.

Greenwell, J. Richard. 1992. Personal communication, July 7.

Heuvelmans, Bernard. 1959. *On the Track of Unknown Animals.* New York: Hill and Wang.

Hills, Daphne M. (Mammal curator, British Museum.) 1994. Personal communication, July 22.

Shuker, Karl P. N. 1994. Personal communication, September 7.

Shuker, Karl P. N. 1993. Personal communication, August 23.

MacFarlane's Bear

Goodwin, George C. 1946. "Inopinatus the Unexpected," *Natural History,* November.

Halfpenny, James. 1995. Personal communication, April 30.

Merriam, C. Hart. 1918. "Vetularctos, A New Genus Related to Ursus," *North American Fauna,* Vol. 41.

Giant Eel

Anonymous. 1971. "Giant *Leptocephalus,*" *Nature,* April 2.

Heuvelmans, Bernard. 1968. *In the Wake of the Sea-Serpents.* New York: Hill and Wang.

Ley, Willy. 1959. *Willy Ley's Exotic Zoology.* New York: Viking.

Shuker, Karl P. N. 1993. *The Lost Ark.* London: HarperCollins.

Albatross and *Deepstar* Reports

Editors of Reader's Digest. 1986. *Sharks: Silent Hunters of the Deep.* New York: Reader's Digest Books.

Kaharl, Victoria A. 1990. *Water Baby: the Story of Alvin.* New York: Oxford University Press.

Lineaweaver, Thomas H., and Richard H. Backus. 1970. *The Natural History of Sharks.* Philadelphia: J. B. Lippincott.

McCormick, Harold W., et. al. 1978. *Shadows in the Sea: the Sharks, Skates and Rays*. New York: Stein and Day.

Ricciuti, Edward. 1973. *Killers of the Seas*. New York: Collier Books.

Soule, Gardner. 1968. *Undersea Frontiers*. Rand McNally, Chicago.

Soule, Gardner. 1970. *Wide Ocean*. Rand McNally, Chicago.

Wood, Gerald L. 1977. *Animal Facts and Feats*. Sterling Publishing Co., New York.

Pygmy Elephant

Anonymous. 1991. "I Dig a Pygmy," *OMNI*, June, quoting *Journal of the Cologne Zoo*, Vol. 32, No. 2.

Balouet, Jean-Christophe, and Eric Alibert. 1990. *Extinct Species of the World*. New York: Barron's.

Carrington, Richard. *Elephants*. 1959. New York: Basic Books.

Chadwick, Douglas H. 1992. *The Fate of the Elephant*. San Francisco: Sierra Club Books.

Dixson, A. F. 1981. *The Natural History of the Gorilla*. New York: Columbia University Press.

Matthiessen, Peter. 1991. *African Silences*. New York: Random House.

Orenstein, Richard (ed.). 1991. *Elephants: The Deciding Decade*. San Francisco: Sierra Club Books.

Shuker, Karl P. N. 1993. *The Lost Ark*. London: HarperCollins.

Wood, Gerald L. 1977. *Animal Facts and Feats*. New York: Sterling Publishing Co.

Truckfish and Checkerboard Fish

Cousteau, Jacques. 1963. *The Living Sea*. New York: Harper & Row.

Grzimek, Bernhard. 1974. *Grzimek's Animal Life Encyclopedia*, Vol. 5. New York: Van Nostrand Reinhold Company.

Soule, Gardner. 1968. *Undersea Frontiers*. Rand McNally, Chicago.

Imitation Sea Monster

Cohen, Daniel. 1971. "Sea Serpents: What they Really Are," *Science Digest*, March.

"Lucio," 1942. Poem in the *Manchester Guardian Weekly*, February 13.

Kamchatka Bear

Day, David. 1990. *The Doomsday Book of Animals*. New York: Viking Press.

Domico, Terry. 1988. *Bears of the World*. New York: Facts on File.

Wood, Gerald L. 1977. *Animal Facts and Feats*. New York: Sterling Publishing Co.

Freaks, Hybrids, and Fakes

Anonymous. 1968. "The art of standing out in a crowd, coming naturally to a spotted zebra," London *Daily Mirror*, January 3.

Baker, Mary L. 1987. *Whales, Dolphins, and Porpoises of the World*. Garden City, New York: Doubleday & Co.

Bueler, Lois E. 1973. *Wild Dogs of the World*. New York: Stein and Day.

Cousins, Don. 1982. "Ape Mystery," *BBC Wildlife*, April.

Fraser, F.C. 1966. "Comments on the Delphinoidea." In Norris, K. S. *Whales, Dolphins, and Porpoises*. Berkeley, CA: University of California Press.

183

Guggisberg, C.A.W. 1975. *Wild Cats of the World*. New York: Taplinger Publishing Co.

Ley, Willy. 1959. *Willy Ley's Exotic Zoology*. New York: Viking Press.

McCormick, Harold W., et. al. 1978. *Shadows in the Sea: the Sharks, Skates and Rays*. New York: Stein and Day.

Ryder, Oliver A. 1994. "A Horse of a Different Chromosome?" *Natural History*, October.

Shuker, Karl P.N. 1989. *Extraordinary Animals Worldwide*. London: Robert Hale.

Giant Octopus

Bright, Michael. 1989. *There are Giants in the Sea*. London: Robson Books.

Ellis, Richard. 1994. *Monsters of the Sea*. New York: Alfred A. Knopf.

Holden, Constance (ed.). 1995. "One Sea Monster Down," *Science*, April 14.

Mackal, Roy. 1986. "Biochemical Analysis of Preserved *Octopus Giganteus* Tissue," *Cryptozoology*, Vol. 5.

Raynal, Michel. 1991. "Le 'Monster de Floride' de 1896: Cetace ou Poulpe Colossal?" *Bulletin de Societe neuchateloise*, no. 114.

Wood, F. G. 1971. "Octopus Trilogy," *Natural History*, March.

Wood, Gerald L. 1977. *Animal Facts and Feats*. New York: Sterling Publishing Co.

Kon-Tiki Reports

Heyerdahl, Thor. 1950. *Kon-Tiki*. Chicago: Rand McNally.

King Cheetah

Bottriel, Lena Godsall. 1987. *King Cheetah*. New York: E. J. Brill.

George, Jean. 1982. "The King Cheetah Puzzle," *BBC Wildlife*, February.

Sayre, Roxanna. 1984. "Creatures," *Audubon*, May.

van Aarde, R. J., and Ann van Dyk. 1986. "Inheritance of the king coat color pattern in cheetahs *Acinonyx jubatus*," *Journal of Zoology* (London).

Giant Water Hen

Fuller, Errol. 1987. *Extinct Birds*. New York: Facts on File.

Mackal, Roy. 1980. *Searching for Hidden Animals*. New York: Doubleday & Co.

Tunicate

Kaharl, Victoria A. 1990. *Water Baby: the Story of Alvin*. New York: Oxford University Press.

Mackal, Roy. 1980. *Searching for Hidden Animals*. New York: Doubleday & Co.

Soule, Gardner. 1968. *Undersea Frontiers*. Rand McNally, Chicago.

Kellas Cat

Shuker, Karl P. N. 1993. Personal communication, August 23.

Shuker, Karl P. N. 1989. *Mystery Cats of the World*. London: Robert Hale.

Great White and Giant Squid

Bright, Michael. 1989. *There are Giants in the Sea*. London: Robson Books.

Ellis, Richard, and John E. McCosker. 1991. *Great White Shark*. New York: HarperCollins.

Ellis, Richard. 1976. *The Book of Sharks*. New York: Grosset & Dunlap. Also 1983 edition published by Alfred A. Knopf.

Steel, Rodney. 1985. *Sharks of the World*. New York: Facts on File.

Watson, Lyall. 1981. *Sea Guide to Whales of the World*. New York: E.P. Dutton.

Double-Banded Argus
Carlson, Carl W. 1981. "The Sutton's Warbler - A Critical Review and Summation of Current Data," *Atlantic Naturalist,* Vol. 34.

Krumbiegel, Ingo. 1986. "The Unseen Argus Pheasant," *Cryptozoology,* Vol. 5.

Fuller, Errol. 1987. *Extinct Birds.* New York: Facts on File.

Seafloor Trails
Heezen, Brian, and Charles Hollister. 1971. *The Face of the Deep.* New York: Oxford University Press.

Mackal, Roy. 1980. *Searching for Hidden Animals.* New York: Doubleday & Co.

Giant Gecko
Bauer, Anthony M., and Anthony P. Russell. 1988. "Osteological Evidence for the Prior Occurrence of a Giant Gecko in Otago, New Zealand," *Cryptozoology,* Vol. 7.

Mystery Whales
Bright, Michael. 1989. *There are Giants in the Sea.* London: Robson Books.

Bruyuns, W. F. J. Morzer. 1971. *Field Guide of Whales and Dolphins.* Amsterdam: C.A. Mees.

Ellis, Richard. 1982. *Dolphins and Porpoises.* New York: Alfred A. Knopf.

Heuvelmans, Bernard. 1986. "Annotated Checklist of Apparently Unknown Animals With Which Cryptozoology is Concerned," Cryptozoology, Vol. 5.

Heyning, John E., and Marilyn E. Dahlheim. 1988. "Orcinus Orca," *Mammalian Species,* # 304, American Society of Mammologists, January 15.

Nowack, Ronald M. 1991. *Walker's Mammals of the World.* Baltimore: Johns Hopkins University Press.

Raynal, Michel. 1993. Personal communication, February 5.

Raynal, Michel, and Jean-Pierre Sylvestre. 1991. "Cetaceans with two dorsal fins," *Aquatic Mammals,* Vol. 17.1.

Raynal, Michel. 1987. "The Linnaeus of the Zoology of Tomorrow," *Cryptozoology,* Vol. 6, quoting Berzin, A. A., and Vladimirov, V. L. 1983. "A New Species of Killer Whale from the Antarctic Waters," *Zoologisheski Zhurnal,* Vol. 62(2).

Giant Salamander
Myers, George S. 1951. "Asiatic Giant Salamander Caught in the Sacramento River and an Exotic Skink Near San Francisco," *Copeia,* No. 2.

Rodgers, Thomas L. 1962. "Report of Giant Salamanders in California," *Copeia,* No. 3.

Long-Snouted Moth
Angier, Natalie. 1992. "May Be Elusive but Moth With 15-Inch Tongue Should Be Out There," *New York Times.* January 14.

Huyghe, Patrick. 1993. "New-Species Fever," *Audubon,* March-April.

Maybe Animals
Bright, Michael. 1987. *The Living World.* New York: St. Martin's Press.

Burton, Maurice. 1952. "The Supposed 'Tiger-Cat' of Queensland," *Oryx.*

East, Rod. 1992. "Conservation Status of Antelopes in Asia and the Middle East, Part 1," *Species,* December.

Heuvelmans, Bernard. 1986. "Annotated Checklist of Apparently Unknown Animals with which Cryptozoology is Concerned," *Cryptozoology,* Vol. 5.

Heuvelmans, Bernard. 1968. *In the Wake of the Sea-Serpents.* New York: Hill and Wang.

Hichens, William. 1937. "African Mystery Beasts," *Discovery,* December.

Hutchins, Michael, and Barbara Sleeper. 1993. "Out on a Limb," *Animals*, November/December.

Ley, Willy. 1959. *Willy Ley's Exotic Zoology.* New York: Viking Press.

Mackal, Roy. 1987. *A Living Dinosaur? In Search of Mokele-Mbembe.* New York: E. J. Brill.

Mackal, Roy. 1980. *Searching for Hidden Animals.* New York: Doubleday & Co.

Neill, Wilfred T. 1971. *The Last of the Ruling Reptiles.* New York: Columbia University Press.

Sanderson, Ivan T. 1937. *Animal Treasure.* New York: Viking Press.

Shuker, Karl P. N. 1993. *The Lost Ark.* London: HarperCollins.

Shuker, Karl P.N. 1989. *Extraordinary Animals Worldwide.* London: Robert Hale.

Shuker, Karl P.N. 1989. *Mystery Cats of the World.* London: Robert Hale.

Literary Quotations

Blake, William. "The Tiger," in Williams, Oscar (ed.) 1952. *Immortal Poems of the English Language.* New York: Washington Square Press, Inc.

Lawrence, D.H. 1929. *Birds, Beasts and Flowers.* Reprinted 1992. Santa Rosa, CA: Black Sparrow Press.

Tennyson, Alfred. 1897. *The Poetical Works of Alfred, Lord Tennyson.* New York: Thomas Y. Crowell.

"The Coelacanth," by Ogden Nash, is from *The Old Dog Barks Backwards,* copyright 1972. Quoted by permission of Little, Brown and Company, Boston, Mass.

Index

192